Student Solutions Manual

to accompany

UNIVERSITY PHYSICS

Standard Version

Jeff Sanny & William Moebs
Loyola Marymount University

Prepared by

Bo Lou
Ferris State

Thurman R. Kremser
Albright College

Robert K. Cole
University of Southern California

WCB **Wm. C. Brown Publishers**

Dubuque, IA   Bogota   Boston   Buenos Aires   Caracas   Chicago
Guilford, CT   London   Madrid   Mexico City   Sydney   Toronto

A Times Mirror Company

ISBN 0-697-23258-1

Printed in the United States of America by Times Mirror Higher Education Group, Inc., 2460 Kerper Boulevard, Dubuque, Iowa, 52001

10 9 8 7 6 5 4 3 2 1

PREFACE

This student's solutions manual is written to accompany "**UNIVERSITY PHYSICS**" by Jeff Sanny and William Moebs. It provides detailed solutions to every third-odd (1, 7, 13, 19, 25, 31, etc.) end-of-chapter problem.

Three Professors contributed to this manual: Bo Lou of Ferris State University (Chapter 1 through 10, Chapter 21 through 30), Thurman Kremser of Albright College (Chapter 11 through 20), and Robert Cole of University of Southern California (Chapter 31 through 39).

Correctness and accuracy of the solutions and answers are of paramount importance to both instructors and students. Every possible effort has been taken to ensure an error free and accurate solutions manual. Every problem has been worked out and checked by the authors of this manual, authors of the text book, and other Physics professors.

Bo Lou typeset the manual and drafted the artworks with Microsoft Word for Windows Version 6.0.

TABLE OF CONTENTS

Chapter	Page

Chapter 1 Introduction

1.   b is $\boxed{\text{Length}}$. Its units are $\boxed{\text{meters in SI and centimeters in cgs}}$.

    m is $\boxed{\dfrac{\text{Length}}{\text{Time}}}$. Its units are $\boxed{\text{m/s in SI and cm/s in cgs}}$.

7. $1 \text{ ly} = (3.00 \times 10^8 \text{ m/s})(3.16 \times 10^7 \text{ s}) = \boxed{9.48 \times 10^{15} \text{ m}}$.

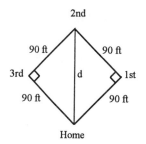

13. $d = \sqrt{(90 \text{ ft})^2 + (90 \text{ ft})^2} = 127 \text{ ft}$

      $= (127 \text{ ft}) \times (\dfrac{0.3048 \text{ m}}{1 \text{ ft}}) = \boxed{39 \text{ m}}$.

19. $1.0 \text{ g/cm}^3 = (1.0 \text{ g/cm}^3) \times (\dfrac{1 \text{ kg}}{1000 \text{ g}}) \times (\dfrac{100 \text{ cm}}{1 \text{ m}})^3 = 1000 \text{ kg/m}^3$,

      $\rho = \dfrac{m}{V}$, $\Rightarrow$ $V = \dfrac{m}{\rho} = \dfrac{70 \text{ kg}}{1000 \text{ kg/m}^3} = \boxed{0.070 \text{ m}^3}$.

25.   Leave to reader.

31. $V = \dfrac{4\pi r^3}{3} = \dfrac{4\pi (0.0703 \text{ m})^3}{3} = \boxed{1.46 \times 10^{-3} \text{ m}^3}$.

Chapter 2 Vectors

1.   (a) $\mathbf{A} + \mathbf{B} = \boxed{15 \text{ at } 38°}$,   (b) $\mathbf{B} + \mathbf{C} = \boxed{11 \text{ at } -35°}$,

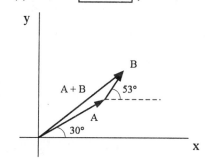

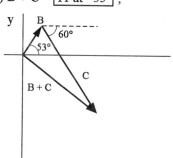

(c) **D** + **E** = $\boxed{24 \text{ at } 161°}$,

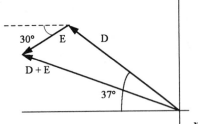

(d) **A** − **B** = $\boxed{5.7 \text{ at } 10°}$,

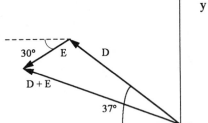

(e) **D** − **E** = $\boxed{18 \text{ at } 120°}$,

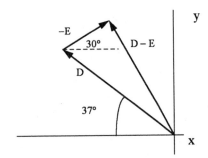

(f) **A** + 2**E** = $\boxed{6.0 \text{ at } 210°}$,

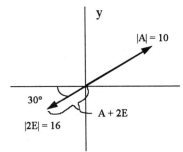

(g) **C** − 2**D** + 3**E** = $\boxed{49 \text{ at } -70°}$,

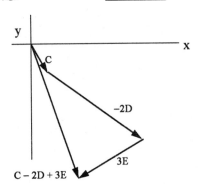

(h) **A** − 4**D** + 2**E** = $\boxed{78 \text{ at } -41°}$.

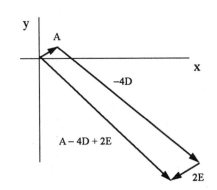

7.   Maximum = a + a + a + a = $\boxed{4a}$,

when all four vectors are parallel.

Minimum = a − a + a − a = $\boxed{0}$,   when every two vectors are anti-parallel.

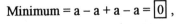

13.   (a) **d** = **d₁** + **d₂** + **d₃** = $\boxed{(-3.0 \, \mathbf{i} - 7.0 \, \mathbf{j} + 3.0 \, \mathbf{k}) \text{ cm}}$,

(b) d = $\sqrt{(-3.0)^2 + (-7.0)^2 + 3.0^2}$ cm = $\boxed{8.2 \text{ cm}}$,

(c) $d_1 = \sqrt{3.0^2 + (-4.0)^2 + (-2.0)^2}$ cm = 5.39 cm , $d_2 = \sqrt{1.0^2 + (-7.0)^2 + 4.0^2}$ cm = 8.12 cm ,

$d_3 = \sqrt{(-7.0)^2 + 4.0^2 + 1.0^2}$ cm = 8.12 cm ,   Distance = $d_1 + d_2 + d_3$ = $\boxed{22 \text{ cm}}$.

19. (a) $\mathbf{V_1} = 1.0\ \mathbf{i}\ m$, $\mathbf{V_2} = (1.0\ m)(\cos 30°\ \mathbf{i} + \sin 30°\ \mathbf{j})$,

$\mathbf{V_3} = (1.0\ m)(\cos 60°\ \mathbf{i} + \sin 60°\ \mathbf{j})$, $\mathbf{V_4} = 1.0\ \mathbf{j}\ m$,

$\mathbf{V_1} + \mathbf{V_2} + \mathbf{V_3} + \mathbf{V_4} = (2.37\ m)(\mathbf{i} + \mathbf{j})$,

Magnitude $= 2.37\sqrt{2}\ m = \boxed{3.4\ m}$, $\theta = \boxed{45°}$,

(b) $\mathbf{V} = \boxed{3.4\ m\ at\ 45°}$.

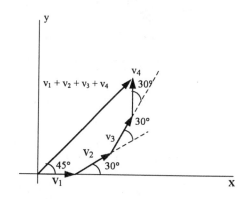

25. $\mathbf{M} \cdot \mathbf{N} = MN\cos\theta = M_x N_x + M_y N_y + M_z N_z$

$= (2.0)(3.0) + (-4.0)(4.0) + (1.0)(10.0) = 0$, So, $\theta = \boxed{90°}$,

31. $P = Q = \sqrt{3.0^2 + 4.0^2} = 5.0$, $\mathbf{P} \cdot \mathbf{Q} = PQ\cos 90° = 0 = M_x N_x + M_y N_y = P_x(3.0) + P_y(4.0)$,

$|\mathbf{P} \times \mathbf{Q}| = PQ\sin 90° = PQ = P_x Q_y - P_y Q_x = P_x(4.0) - P_y(3.0) = (5.0)^2$,

Solve for $P_x = 4.0$, $P_y = -3.0$, So, $\mathbf{P} = \boxed{4.0\ \mathbf{i} - 3.0\ \mathbf{j}}$.

37. $\mathbf{C} = \mathbf{A} + \mathbf{B}$, $\mathbf{C'} = \frac{1}{2}\mathbf{A} + \frac{1}{2}\mathbf{B} = \frac{1}{2}(\mathbf{A} + \mathbf{B}) = \frac{1}{2}\mathbf{C}$, So, $\boxed{\mathbf{C'} = \frac{1}{2}\mathbf{C}}$.

Chapter 3 One-Dimensional Motion

1. (a) $\mathbf{x_1} = \boxed{5.0\ \mathbf{i}\ m}$, (b) $\mathbf{x_2} = \boxed{-3.0\ \mathbf{i}\ m}$.

7. The second-place finisher's speed $v_2 = \dfrac{\Delta d}{\Delta t} = \dfrac{100\ m}{11.6\ s} = 8.62\ m/s$,

After $11.2\ s$, he traveled $(8.62\ m/s)(11.2\ s) = 96.5\ m$. So he is $100\ m - 96.5\ m = \boxed{3.5\ m\ behind}$.

13. $v(t) = \dfrac{dx(t)}{dt} = -8.0t\ m/s$, $v(2.0) = \boxed{-16\ m/s}$, $v(5.0) = \boxed{-40\ m/s}$,

$a(t) = \dfrac{dv(t)}{dt} = \boxed{-8.0\ m/s^2 = constant}$.

19. (a) Acceleration equals the slope of curve in v vs. t graph.

$a = \dfrac{\Delta v}{\Delta t}$, $a(10) = \boxed{0}$, $a(30) = \dfrac{15\ m/s - 5\ m/s}{40\ s - 20\ s} = \boxed{0.50\ m/s^2}$,

$$a(50) = \frac{0 - 15 \text{ m/s}}{50 \text{ s} - 40 \text{ s}} = \boxed{-1.5 \text{ m/s}^2} \ ,$$

(b) Displacement equals the area under curve in **v** vs. t graph.

$$\Delta x(0 \to 40) = (5 \text{ m/s})(20 \text{ s}) + \frac{5 \text{ m/s} + 15 \text{ m/s}}{2} \times (20 \text{ s}) = \boxed{300 \text{ m}} \ ,$$

$$\Delta x(40 \to 60) = \frac{1}{2}(10 \text{ s})(15 \text{ m/s}) - \frac{1}{2}(10 \text{ s})(15 \text{ m/s}) = \boxed{0} \ .$$

25. (a) $x - x_0 = v_0 t + \frac{1}{2}at^2 = (30 \text{ m/s})(5.0 \text{ s}) + \frac{1}{2}(30 \text{ m/s}^2)(5.0 \text{ s})^2 = \boxed{530 \text{ m}} \ ,$

(b) $v = v_0 + at = 30 \text{ m/s} + (30 \text{ m/s}^2)(5.0 \text{ s}) = \boxed{180 \text{ m/s}} \ .$

31. Average $v = \frac{v_0 + v}{2} = \frac{30 \text{ mi/hr} + 0}{2} = \boxed{15 \text{ mi/hr}} \ .$

37. (a) The final velocity of the first mile is the initial velocity of the second mile. Use $x - x_0 = v_0 t + \frac{1}{2}at^2$,

For the first mile, $\quad 1609 \text{ m} = v_0(80 \text{ s}) + \frac{1}{2}a(80 \text{ s})^2 \ , \quad v_1 = v_0 + at = v_0 + a(80 \text{ s}) = v_{20} \ ,$

For the second mile, $\quad 1609 \text{ m} = [v_0 + a(80 \text{ s})](120 \text{ s}) + \frac{1}{2}a(120 \text{ s})^2 \ ,$

Solve for $\quad a = \boxed{-0.067 \text{ m/s}^2} \ , \quad$ and $\quad v_0 = 22.8 \text{ m/s} \ ,$

(b) As in (a) , $\quad v_0 = \boxed{23 \text{ m/s}} \ , \quad v_2 = 22.8 \text{ m/s} + (-0.067 \text{ m/s}^2)(200 \text{ s}) = \boxed{9.4 \text{ m/s}} \ .$

For the fly, $\quad x - x_0 = 0 + \frac{1}{2}(0.10 \text{ m/s}^2)(0.707 \text{ s})^2 = 0.025 \text{ m} < 0.050 \text{ m} \ .$ So the answer is $\boxed{\text{No}}$.

43. (a) $x - x_0 = x = v_0 t - \frac{1}{2}at^2 \ , \quad\quad \frac{dx}{dt} = v_0 - at = 0 \ , \quad\quad t = \boxed{\frac{v_0}{a}} \ .$ The particle stops at this time.

(b) $x = v_0(\frac{v_0}{a}) - \frac{1}{2}a(\frac{v_0}{a})^2 = \boxed{\frac{v_0^2}{2a}} \ .$

49. Choose upward as positive.

(a) $x - x_0 = v_0 t + \frac{1}{2}at^2 \ , \quad \Rightarrow \quad 0 = v_0(4.0 \text{ s}) - \frac{1}{2}(9.80 \text{ m/s}^2)(4.0 \text{ s})^2 \ , \quad\quad v_0 = \boxed{20 \text{ m/s}} \ ,$

(b) $x - x_0 = (19.6 \text{ m/s})(2.0 \text{ s}) - \frac{1}{2}(9.80 \text{ m/s}^2)(2.0 \text{ s})^2 = \boxed{20 \text{ m}} \ .$

55. Choose upward as positive.

(a) Consider only the rising motion.

$v^2 = v_0^2 + 2a(x - x_0) \ , \quad \Rightarrow \quad 0 = v_0^2 - 2(9.80 \text{ m/s}^2)(59 \text{ m}) \ , \quad v_0 = \boxed{34 \text{ m/s}} \ ,$

(b) $x - x_0 = v_0 t + \frac{1}{2}at^2 \ , \quad \Rightarrow \quad -(h + 1.0 \text{ m}) = (34 \text{ m/s})(10 \text{ s}) - \frac{1}{2}(9.80 \text{ m/s}^2)(10 \text{ s})^2 = -150 \text{ m} \ ,$

So, $\quad h = \boxed{149 \text{ m}} \ ,$

(c) $v = v_0 + at = 34$ m/s $- (9.80$ m/s$^2)(10$ s$) = \boxed{-64 \text{ m/s}}$. The $-$ sign means downward.

(d) $\boxed{9.80 \text{ m/s}^2 \text{ downward}}$.

61. (a) $a(t) = \dfrac{dv(t)}{dt}$ $= \boxed{3.2 \text{ m/s}^2 \qquad (0 \le t < 5.0 \text{ s})}$,

$= \boxed{-1.5 \text{ m/s}^2 \qquad (5.0 \text{ s} \le t < 11.0 \text{ s})}$,

$= \boxed{0 \qquad\qquad (t \ge 11.0 \text{ s})}$,

(b) $x(t) = \displaystyle\int v(t)dt$, During $0 \le t < 5.0$ s, $x(t) = \displaystyle\int (3.2t \text{ m/s})dt$ $= 1.6t^2$ m $+ C_1$,

$x(0) = 0$, So, $C_1 = 0$, Therefore $x(t) = 1.6t^2$ m $(0 \le t < 5.0 \text{ s})$,

$x(2.0) = 1.6(2.0)^2$ m $= \boxed{6.4 \text{ m}}$, $x(5.0) = 1.6(5.0)^2$ m $= 40$ m ,

During 5.0 s $\le t < 11.0$ s, $x(t) = \displaystyle\int \{[16.0 - 1.5(t - 5.0)] \text{ m/s}\}dt$ $= 16.0t - \dfrac{1.5(t - 5.0)^2}{2} + C_2$,

$x(5.0) = 40$, So, $C_2 = -40$ m , Therefore $x(t) = [16.0t - 0.75(t - 5.0)^2 - 40]$ m ,

$x(7.0) = [16.0(7.0) - 0.75(7.0 - 5.0)^2 - 40]$ m $= \boxed{69 \text{ m}}$,

$x(11.0) = [16.0(11.0) - 0.75(11.0 - 5.0)^2 - 40]$ m $= 109$ m ,

$x(12.0) = 109$ m $+ (7.0 \text{ m/s})(1.0 \text{ s}) = \boxed{116 \text{ m}}$.

67. Choose downward as positive. The final velocity from release to top of the window is the initial velocity passing the window.

For passing the window, $x - x_0 = v_0 t + \dfrac{1}{2}at^2$, $\Rightarrow$ 3.0 m $= v_0(0.20$ s$) + \dfrac{1}{2}(9.80$ m/s$^2)(0.20$ s$)^2$,

$v_0 = 14$ m/s , From release to top of the window, $v^2 = v_0{}^2 + 2a(x - x_0)$,

So, $(14 \text{ m/s})^2 = 0 + 2(9.80 \text{ m/s}^2)(x - x_0)$, $x - x_0 = \boxed{10 \text{ m}}$.

Chapter 4 Motion in Two and Three Dimensions

1. $\mathbf{r} = \boxed{(1.0 \text{ } \mathbf{i} - 4.0 \text{ } \mathbf{j} + 6.0 \text{ } \mathbf{k}) \text{ m}}$.

7. $\mathbf{v} = \dfrac{\Delta \mathbf{r}}{\Delta t} = \dfrac{-6.0 \text{ } \mathbf{i} \text{ cm}}{0.20 \text{ s}} = \boxed{-30 \text{ } \mathbf{i} \text{ cm/s}}$.

13.   (a) $\mathbf{v}(t) = \dfrac{d\mathbf{r}(t)}{dt} = c_1 \mathbf{i} - 2c_2 t \mathbf{j} = \boxed{50\ \mathbf{i}\ \text{m/s} - (9.8\ \text{m/s}^2)t\ \mathbf{j}}$,

        $\mathbf{a}(t) = \dfrac{d\mathbf{v}(t)}{dt} = -2c_2 \mathbf{j} = \boxed{-9.8\ \mathbf{j}\ \text{m/s}^2}$,

        (b) Initial velocity $\mathbf{v_0} = \boxed{50\ \mathbf{i}\ \text{m/s}}$,

        Initial position $\mathbf{r_0} = \boxed{0}$,

        Initial acceleration $\mathbf{a_0} = \boxed{-9.8\ \mathbf{j}\ \text{m/s}^2}$.

19. $\mathbf{v}(t) = \mathbf{v_o} + \mathbf{a}t = 8.0\ \mathbf{j}\ \text{m/s} + \left[(4.0\ \mathbf{i} + 3.0\ \mathbf{j})\ \text{m/s}^2\right]t = \boxed{[4.0t\ \mathbf{i} + (3.0t + 8.0)\ \mathbf{j}]\ \text{m/s}}$,

        $\mathbf{r}(t) = \mathbf{r_o} + \mathbf{v_o}t + \dfrac{1}{2}\mathbf{a}t^2 = \left[(5.0\ \mathbf{i} + 2.0\ \mathbf{j})\ \text{m}\right] + (8.0\ \mathbf{j}\ \text{m/s})t + \dfrac{1}{2}\left[(4.0\ \mathbf{i} + 3.0\ \mathbf{j})\ \text{m/s}^2\right]t^2$

        $= \boxed{[(2.0t^2 + 5.0)\ \mathbf{i} + (1.5t^2 + 8.0t + 2.0)\ \mathbf{j}]\ \text{m}}$,

25.   Choose upward as positive.

        (a) $y - y_0 = v_{oy} - \dfrac{1}{2}gt^2$, $\Rightarrow$ $-1.0\ \text{m} = 0 - \dfrac{1}{2}(9.80\ \text{m/s}^2)t^2$, $t = \boxed{0.45\ \text{s}}$,

        (b) $x - x_0 = v_{ox}t$, $\Rightarrow$ $3.0\ \text{m} = v_{ox}(0.452\ \text{s})$, $v = v_{ox} = \boxed{6.6\ \text{m/s}}$,

        (c) $v_y = v_{oy} - gt = 0 - (9.80\ \text{m/s}^2)(0.452\ \text{s}) = -4.43\ \text{m/s}$,   Speed $= \sqrt{6.64^2 + (-4.43)^2}\ \text{m/s} = \boxed{8.0\ \text{m/s}}$.

31.   Choose upward as positive.

        (a) $v_{ox} = 90\ \text{mi/hr} = 132\ \text{ft/s}$, $x - x_0 = v_{ox}t$, $\Rightarrow$ $55\ \text{ft} = (132\ \text{ft/s})t$, $t = \boxed{0.42\ \text{s}}$,

        (b) $y - y_0 = v_{oy}t - \dfrac{1}{2}gt^2 = 0 - \dfrac{1}{2}(32\ \text{ft/s}^2)(0.417\ \text{s})^2 = \boxed{-2.8\ \text{ft}}$.

37. $R = \dfrac{v_0^2 \sin 2\theta}{g}$,   So, $\dfrac{R_{30}}{R_{45}} = \dfrac{\sin 60°}{\sin 90°} = 0.866$,

        So the distance lost $= (1 - 0.866)R_{45} = 0.134(8.0\ \text{m}) = \boxed{1.1\ \text{m}}$.

43. $R = \dfrac{v_0^2 \sin 2\theta}{g}$, $\Rightarrow$ $g = \dfrac{v_0^2 \sin 2\theta}{R} = \dfrac{(10\ \text{m/s})^2 \sin 90°}{20\ \text{m}} = \boxed{5.0\ \text{m/s}^2}$,

        $R' = \dfrac{(10\ \text{m/s})^2 \sin 120°}{5.0\ \text{m/s}^2} = \boxed{17\ \text{m}}$.

49. $v = \dfrac{2\pi r}{T} = \dfrac{2\pi r}{60\ \text{s}}$, $a = 0.10\ \text{cm/s}^2 = \dfrac{v^2}{r} = \dfrac{4\pi^2 r^2}{(60\ \text{s})^2 r} = \dfrac{4\pi^2 r}{(60\ \text{s})^2}$, $r = \boxed{9.1\ \text{cm}}$.

55.   Use the following subscripts: b = boat,   w = water,   s = shore. $\mathbf{v_{bs}} = \mathbf{v_{bw}} + \mathbf{v_{ws}}$,

        (a) $v_{bs} = 8.0\ \text{km/hr} + 3.0\ \text{km/hr} = 11.0\ \text{km/hr}$, $\Delta t = \dfrac{\Delta d}{v} = \dfrac{1.5\ \text{km}}{11.0\ \text{km/hr}} = 0.136\ \text{hr} = \boxed{8.2\ \text{min}}$,

(b) $v_{bs} = 8.0 \text{ km/hr} - 3.0 \text{ km/hr} = 5.0 \text{ km/hr}$, $\Delta t = \dfrac{1.5 \text{ km}}{5.0 \text{ km/hr}} = 0.30 \text{ hr} = \boxed{18 \text{ min}}$,

(c) $\theta = \sin^{-1} (\dfrac{3.0}{8.0}) = \boxed{22°}$,

(d) $v_{bs} = \sqrt{8.0^2 - 3.0^2} \text{ km/hr} = \boxed{7.4 \text{ km/hr}}$, $\Delta t = \dfrac{0.80 \text{ km}}{7.42 \text{ km/hr}} = 0.108 \text{ hr} = \boxed{6.5 \text{ min}}$,

(e) $\Delta t = \dfrac{0.80 \text{ km}}{8.0 \text{ km/hr}} = 0.10 \text{ hr} = \boxed{6.0 \text{ min}}$, $\Delta d = v_{ws}t = (3.0 \text{ km/hr})(0.10 \text{ hr}) = \boxed{0.30 \text{ km}}$.

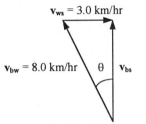
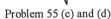

Problem 55 (c) and (d)

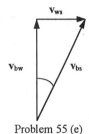

Problem 55 (e)

61. $v_0 = 60 \text{ km/hr} = 16.7 \text{ m/s}$, $y - y_0 = v_{oy}t - \dfrac{1}{2}gt^2$,

So, $-100 \text{ m} = [-(16.7 \text{ m/s})\sin 30°]t - \dfrac{1}{2}(9.80 \text{ m/s}^2)t^2$, $t = 3.75 \text{ s}$,

$x - x_0 = v_0 \cos\theta t = (16.7 \text{ m/s})\cos 30°(3.75 \text{ s}) = 54 \text{ m} < 60 \text{ m}$. $\boxed{\text{He does not make it}}$.

67.   (a) Relative to a reference frame attached to the plane.

For the missile, $v_0 = 150 \text{ m/s} - 100 \text{ m/s} = 50 \text{ m/s}$, $a = 1.0 \text{ m/s}^2 - 2.0 \text{ m/s}^2 = -1.0 \text{ m/s}^2$,

$x - x_0 = v_0 t + \dfrac{1}{2}at^2$, $\Rightarrow$ $200 \text{ m} = (50 \text{ m/s})t + \dfrac{1}{2}(-1.0 \text{ m/s}^2)t^2$, $t = 4.18 \text{ s}$ or 96 s.

$\boxed{\text{Yes, the missile will catch the plane}}$.

(b) $v_0 = 150 \text{ m/s} - 130 \text{ m/s} = 20 \text{ m/s}$, $a = 1.0 \text{ m/s}^2 - 3.0 \text{ m/s}^2 = -2.0 \text{ m/s}^2$,

$200 \text{ m} = (20 \text{ m/s})t + \dfrac{1}{2}(-2.0 \text{ m/s}^2)t^2$,   there is no real solution for t.

$\boxed{\text{No, the missile will not catch the plane}}$.

Chapter 5 Dynamics I

1.   (a) $90 \text{ km/hr} = 25 \text{ m/s}$, $a = \dfrac{\Delta v}{\Delta t} = \dfrac{25 \text{ m/s} - 0}{10 \text{ s}} = \boxed{2.5 \text{ m/s}^2}$,

(b) $\Sigma F = ma = (1000 \text{ kg})(2.5 \text{ m/s}^2) = \boxed{2500 \text{ N}}$.

7. $\Sigma \mathbf{F} = 10.0 \text{ i N} - 2.0 \text{ i N} - 4.0 \text{ j N} = (8.0 \text{ i} - 4.0 \text{ j}) \text{ N}$,

$$\mathbf{a} = \frac{\Sigma \mathbf{F}}{m} = \frac{(8.0 \ \mathbf{i} - 4.0 \ \mathbf{j}) \ N}{5.0 \ kg} = \boxed{(1.6 \ \mathbf{i} - 0.80 \ \mathbf{j}) \ m/s^2} \ .$$

13. $$\mathbf{a} = \frac{\Delta \mathbf{v}}{\Delta t} = \frac{(-2.0 \ \mathbf{i} + 4.0 \ \mathbf{k}) \ m/s - (3.0 \ \mathbf{i} - 6.0 \ \mathbf{j} + 4.0 \ \mathbf{k}) \ m/s}{8.0 \ s - 6.0 \ s} = (-2.5 \ \mathbf{i} + 3.0 \ \mathbf{j}) \ m/s^2 \ ,$$

$$\Sigma \mathbf{F} = m\mathbf{a} = (3.0 \ kg)[(-2.5 \ \mathbf{i} + 3.0 \ \mathbf{j}) \ m/s^2] = \boxed{(-7.5 \ \mathbf{i} + 9.0 \ \mathbf{j}) \ N} \ .$$

19. (a) $W = mg = (15 \ kg)(9.80 \ m/s^2) = \boxed{147 \ N}$.   (b) $W = (15 \ kg)(1.7 \ m/s^2) = \boxed{26 \ N}$.

(c) $m = \boxed{15 \ kg}$,   (d) $W = \boxed{0}$,   (e) $m = \boxed{15 \ kg}$.

25. $$W = F = G \frac{m_1 m_2}{r^2} \ , \qquad \frac{W_p}{W_e} = \frac{m_p}{m_e} = 10 \ , \qquad So, \qquad m_p = \boxed{10 m_e} \ .$$

31. $$\Sigma \mathbf{F} = (6.0 \ \mathbf{i} + 14 \ \mathbf{j}) \ N - (10.0 \ kg)(9.80 \ m/s^2) \ \mathbf{j} = (6.0 \ \mathbf{i} - 84 \ \mathbf{j}) \ N \ ,$$

$$\mathbf{a} = \frac{\Sigma \mathbf{F}}{m} = \frac{(6.0 \ \mathbf{i} - 84 \ \mathbf{j}) \ N}{10.0 \ kg} = \boxed{(0.60 \ \mathbf{i} - 8.4 \ \mathbf{j}) \ m/s^2} \ .$$

37. $$g(r) = a(r) = G \frac{m}{r^2} \ , \qquad \frac{dg(r)}{dr} = -\frac{2Gm}{r^3} \ ,$$

$$\Delta g = -\frac{2Gm}{r^3} \ \Delta r = -\frac{2Gmh}{R_e^3} = -\frac{2h}{R_e} \cdot \frac{Gm}{R_e^2} = \boxed{-\frac{2gh}{R_e}} \ ,$$

$$\Delta g = -\frac{2(9.80 \ m/s^2)(8848 \ m)}{6.37 \times 10^6 \ m} = \boxed{-0.027 \ m/s^2} \ .$$

Chapter 6 Dynamics II

1. $$\Sigma F_y = N - mg - P\sin\theta = ma_y = 0 \ ,$$

$$N = mg + P\sin\theta \ ,$$

$$\Sigma F_x = P\cos\theta - f = ma_x = 0 \ ,$$

$$P\cos\theta - \mu N = P\cos\theta - \mu(mg + P\sin\theta) = 0 \ ,$$

$$So, \qquad P = \frac{\mu mg}{\cos\theta - \mu\sin\theta} \ ,$$

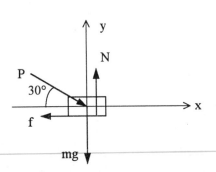

For $\mu_s = 0.70$,

$$P = \frac{(0.70)(10 \ kg)(9.80 \ m/s^2)}{\cos 30° - 0.70\sin 30°} = \boxed{130 \ N} \ ,$$

For $\mu_k = 0.50$,

$$P = \frac{(0.50)(10 \ kg)(9.80 \ m/s^2)}{\cos 30° - 0.50\sin 30°} = \boxed{80 \ N} \ .$$

7. $\Sigma F_y = N - mg = ma_y = 0$, $\qquad$ $N = mg$,

$\Sigma F_x = 10 \text{ N} - f = 10 \text{ N} - \mu N = 10 \text{ N} - \mu mg = ma_x$,

Since $10 \text{ N} > \mu_s mg = 0.80(1.0 \text{ kg})(9.80 \text{ m/s}^2) = 7.84 \text{ N}$, the object is accelerating. Use μ_k.

$a_x = \dfrac{10 \text{ N} - 0.70(1.0 \text{ kg})(9.80 \text{ m/s}^2)}{1.0 \text{ kg}} = 3.14 \text{ m/s}^2$, $\qquad$ $v = v_0 + at = \boxed{(3.14 \text{ m/s}^2)t}$.

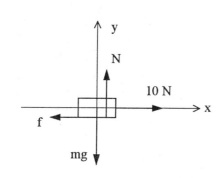

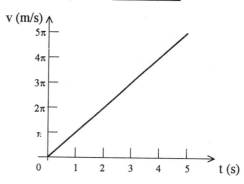

13. (a) $\Sigma F_x = mg\sin\theta = ma_x$,

$a_x = g\sin\theta = (9.80 \text{ m/s}^2)\sin 20° = \boxed{3.4 \text{ m/s}^2}$,

(b) $x - x_0 = v_0 t + \dfrac{1}{2}at^2$,

$30 \text{ m} = 0 + \dfrac{1}{2}(3.35 \text{ m/s}^2)t^2$, $\qquad$ $t = \boxed{4.2 \text{ s}}$.

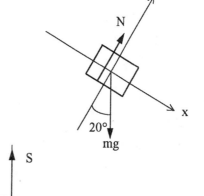

19. (a) $\Sigma F_y = mg - S = m(9.80 \text{ m/s}^2) - 60 \text{ N}$

$= ma_y = m(3.8 \text{ m/s}^2)$, $\qquad$ $m = \boxed{10 \text{ kg}}$.

(b) $S = \boxed{60 \text{ N}}$.

(c) $S = mg = (10 \text{ kg})(9.80 \text{ m/s}^2) = \boxed{98 \text{ N}}$.

(d) $S = \boxed{0}$.

25. For the cab, $\qquad$ $\Sigma F_x = F - T_1 = m_2 a_x$

$= (5000 \text{ kg})(0.50 \text{ m/s}^2) = 2500 \text{ N}$,

For the middle trailer,

$\Sigma F_x = T_1 - T_2 = m_1 a_x$

$= (2000 \text{ kg})(0.50 \text{ m/s}^2) = 1000 \text{ N}$,

For the left trailer,

$\Sigma F_x = T_2 = m_1 a_x$

$= (2000 \text{ kg})(0.50 \text{ m/s}^2) = 1000 \text{ N}$, Solve for

(a) $T_1 = \boxed{2000 \text{ N}}$, $\qquad$ $T_2 = \boxed{1000 \text{ N}}$, $\qquad$ (b) $F = \boxed{4500 \text{ N}}$.

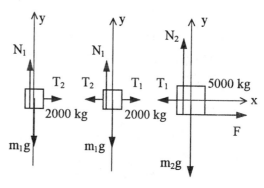

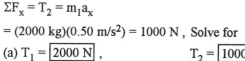

31. For m_2 ,

$\Sigma F_x = m_2 g \sin 53° - T = m_2 a_x$,

$(15 \text{ kg})(9.80 \text{ m/s}^2)\sin 53° - T = (15 \text{ kg})a_x$,

For m_1 ,

$\Sigma F_x = T - m_1 g \sin 37° = m_1 a_x$,

$T - (10 \text{ kg})(9.80 \text{ m/s}^2)\sin 37° = (10 \text{ kg})a_x$,

Solve for $a_x = \boxed{2.3 \text{ m/s}^2}$,

and $T = \boxed{82 \text{ N}}$.

37. $F = -kx$, $\Rightarrow$ $k = -\dfrac{F}{x} = -\dfrac{-(0.25 \text{ kg})(9.80 \text{ m/s}^2)}{0.050 \text{ m}} = 49 \text{ N/m}$,

Spring force provides centripetal force. Assume the spring is stretched by x .

Each revolution is $2\pi r$, So, $v = [2\pi(x + 0.040 \text{ m})/\text{rev}](2.0 \text{ rev/s}) = [4\pi(x + 0.040 \text{ m})] \text{ m/s}$,

$F_r = kx = (49 \text{ N/m})x = ma_r = m\dfrac{v^2}{r} = \dfrac{(0.25 \text{ kg})[4\pi(x + 0.040 \text{ m}) \text{ m/s}]^2}{(x + 0.040 \text{ m})} = [4\pi^2(x + 0.040 \text{ m})] \text{ N}$,

Solve for $x = \boxed{0.17 \text{ m}}$.

43. Static friction provides centripetal force.

$f_s = \mu_s N = \mu_s mg = F_r = ma_r = m\dfrac{v^2}{r}$,

$v = \sqrt{\mu_s gr} = \sqrt{0.70(9.80 \text{ m/s}^2)(65 \text{ m})} = \boxed{21 \text{ m/s}}$.

49. $\dfrac{T^2}{R^3} = \dfrac{4\pi^2}{GM_S}$, Since R and M_S are the same, so is T. $T = \boxed{1 \text{ year}}$.

55. $\Sigma F_y = N - mg\cos 30° - F\sin 30° = ma_y = 0$,

$N = (10 \text{ kg})(9.80 \text{ m/s}^2)\cos 30° + (200 \text{ N})\sin 30° = 185 \text{ N}$,

$\Sigma F_x = F\cos 30° - mg\sin 30° - f_k$

$= F\cos 30° - mg\sin 30° - \mu_k N$

$= (200 \text{ N})\cos 30° - (10 \text{ kg})(9.80 \text{ m/s}^2)\sin 30° - 0.50(185 \text{ N})$

$= 31.7 \text{ N} = ma_x = (10 \text{ kg})a_x$,

So, $a_x = \boxed{3.2 \text{ m/s}^2}$.

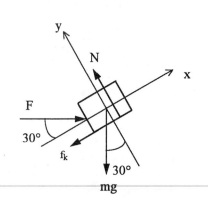

61. (a) Each revolution is $2\pi r$,   So, $v = [2\pi(0.20 \text{ m})\sin\theta/\text{rev}](4.0 \text{ rev/s}) = 5.03\sin\theta \text{ m/s}$,

$\Sigma F_y = N\cos\theta - mg = ma_y = 0$, $N = \dfrac{mg}{\cos\theta}$,

$\Sigma F_x = \Sigma F_r = N\sin\theta = ma_r = m\dfrac{v^2}{r}$,

So, $g\tan\theta = \dfrac{v^2}{r} = \dfrac{(5.03\sin\theta)^2 \text{ m}^2/\text{s}^2}{(0.20 \text{ m})\sin\theta}$,

$\theta = \cos^{-1}\left(\dfrac{9.80}{127}\right) = \boxed{86°}$,   (b) $\boxed{\text{No}}$.

(c) $v = 1.26\sin\theta \text{ m/s}$, $\cos\theta = \dfrac{9.80}{7.94} > 1$,

So the bead $\boxed{\text{sits at the bottom}}$.

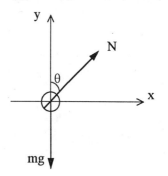

Chapter 7 Work and Mechanical Energy

1. $W = Fd\cos\theta = (20 \text{ N})(5.0 \text{ m})\cos0° = \boxed{100 \text{ J}}$.

7. $W = \displaystyle\int_{2.0 \text{ m}}^{5.0 \text{ m}} F(x)dx = (-2.0)\int_{2.0 \text{ m}}^{5.0 \text{ m}} \dfrac{dx}{x} = (-2.0)\ln x \Big|_{2.0 \text{ m}}^{5.0 \text{ m}} = \boxed{-1.8 \text{ J}}$.

13.   (a) $T = \dfrac{1}{2}mv^2 = \dfrac{1}{2}(0.0080 \text{ kg})(800 \text{ m/s})^2 = \boxed{2.6 \text{ kJ}}$,   (b) $T = \dfrac{1}{2}(0.0080 \text{ kg})(400 \text{ m/s})^2 = \boxed{640 \text{ J}}$.

19. $W = Fd\cos0° = \Delta T = \dfrac{1}{2}m(0)^2 - \dfrac{1}{2}mv^2$,

$F = -\dfrac{mv^2}{2d} = -\dfrac{(0.0080 \text{ kg})(800 \text{ m/s})^2}{2(0.20 \text{ m})} = \boxed{-1.3\times10^4 \text{ N}}$.

25. $E_0 = T_0 + U_0 = \dfrac{1}{2}m(0)^2 + mgh$, $E = T + U = \dfrac{1}{2}mv^2 + mg(0)$,

$\dfrac{E_0}{E} = \dfrac{2gh}{v^2} = \dfrac{2(9.80 \text{ m/s}^2)(600 \text{ m})}{(9.0 \text{ m/s})^2} = \boxed{150}$. $\boxed{\text{Air friction accounts for the loss}}$.

31. $F(x) = -\dfrac{dU(x)}{dx} = \boxed{-\dfrac{a}{x^2} + \dfrac{2b}{x^3}}$.

37. $F = -kx$, $k = -\dfrac{-150 \text{ N}}{0.50 \text{ m}} = 300 \text{ N/m}$, $T + U_s = \text{constant}$.

$$\frac{1}{2}m(0)^2 + \frac{1}{2}kx^2 = \frac{1}{2}mv^2 + \frac{1}{2}k(0)^2 , \qquad \text{So,} \qquad v = \sqrt{\frac{k}{m}}\ x = \sqrt{\frac{300\ \text{N/m}}{0.050\ \text{kg}}}\ (0.50\ \text{m}) = \boxed{39\ \text{m/s}}.$$

43. Choose the bottom of the circle $U = 0$.   Use $T + U = $ constant.

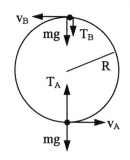

$$\frac{1}{2}mv_A^2 + mg(0) = \frac{1}{2}mv_B^2 + mg(2R) , \qquad v_B^2 = v_A^2 - 4gR ,$$

At the bottom, $\Sigma F_r = T_A - mg = ma_r = m\dfrac{v_A^2}{R}$,

$$T_A = m(\frac{v_A^2}{R} + g) ,$$

At the top, $\Sigma F_r = T_B + mg = m\dfrac{v_B^2}{R} = m\dfrac{v_A^2 - 4gR}{R}$,

$$T_B = m(\frac{v_A^2}{R} - 5g) , \qquad\qquad \text{So,} \qquad T_A - T_B = \boxed{6mg} .$$

49. (a) Choose $U_e(\infty) = U_s(\infty) = 0$.   Use $T + U_e + U_s = $ constant.

$$\frac{1}{2}m(100\ \text{m/s})^2 - \frac{GmM_e}{1.00\times10^8\ \text{m}} - \frac{GmM_s}{1.50\times10^{11} - 1.00\times10^8\ \text{m}}$$

$$= \frac{1}{2}mv^2 - \frac{GmM_e}{6.48\times10^6\ \text{m}} - \frac{GmM_s}{1.50\times10^{11} - 6.48\times10^6\ \text{m}} ,$$

$M_e = 5.98\times10^{24}\ \text{kg}$,   and $M_s = 1.99\times10^{30}\ \text{kg}$,   Solve for $v = \boxed{1.1\times10^4\ \text{m/s}}$,

(b) $\boxed{\text{No, it is not important}}$.

55. Chose the bottom as $U = 0$.

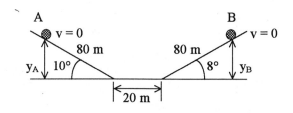

$$W_f = W_{ncon} = \Delta E = \Delta T + \Delta U = \frac{1}{2}m(0^2 - 0^2)$$

$$+ (70\ \text{kg})(9.80\ \text{m/s}^2)(80\ \text{m})(\sin 8° - \sin 10°)$$

$$= \boxed{-1.9\ \text{kJ}} .$$

61. $P = \dfrac{\Delta E}{\Delta t} = \dfrac{1.05\times10^7\ \text{J}}{86400\ \text{s}} = \boxed{122\ \text{W}}$.

67. $13\ \text{km/hr} = 3.61\ \text{m/s}$,   On level surface, $\Sigma F_x = F - f_k = ma_x = 0$, $F = f_k$,

$P = Fv$, $\Rightarrow$ $(0.25)(850\ \text{W}) = f_k(3.61\ \text{m/s})$, $f_k = 58.9\ \text{N}$,

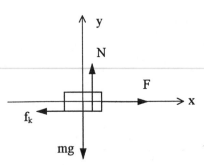

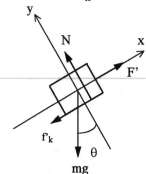

On slope, $\Sigma F_x = F' - f'_k - mg\sin\theta = ma_x = 0$, $F' = f'_k + mg\sin\theta$,

$P' = F'v = [58.9 \text{ N} + (70 \text{ kg})(9.80 \text{ m/s}^2)\sin5.0°](3.61 \text{ m/s}) = 428 \text{ W}$,

Therefore the rate of energy $= \dfrac{428 \text{ W}}{0.25} = \boxed{1.7 \text{ kW}}$.

73. (a) See graph on next page. (b) $F(x) = -\dfrac{dU(x)}{dx} = \boxed{-kx + 2\alpha Axe^{-\alpha x^2}}$,

(c) The energy of the particle at $x = a$ is $E(a) = T(a) + U(a) = \dfrac{1}{2}mv_a^2 + \dfrac{ka^2}{2} + Ae^{-\alpha a^2}$,

The minimum energy needed to reach $x = 0$ is $E(0) = T(0) + U(0) = 0 + A$,

So, $\dfrac{1}{2}mv_a^2 + \dfrac{ka^2}{2} + Ae^{-\alpha a^2} \geq A$, or, $A \leq \boxed{\dfrac{mv_a^2 + ka^2}{2(1 - e^{-\alpha a^2})}}$.

The particle must have enough kinetic energy at the valleys to overcome the potential energy barrier to reach the peak at $x = 0$.

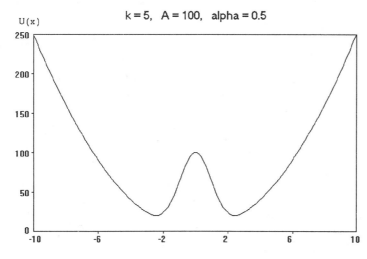

79. (a) Choose where the block momentarily stopped as $U_g = 0$. Use $T + U_g + U_s = \text{constant}$.

$\dfrac{1}{2}m(0)^2 + (2.0 \text{ kg})(9.80 \text{ m/s}^2)(d + 0.20 \text{ m})\sin37° + \dfrac{1}{2}k(0)^2 = \dfrac{1}{2}m(0)^2 + mg(0) + \dfrac{1}{2}(500 \text{ N/m})(0.20 \text{ m})^2$,

Solve for $d = \boxed{0.65 \text{ m}}$,

(b) $W_{sp} + W_{mg} = \dfrac{1}{2}mv^2 = mg(d + x)\sin37° - \dfrac{1}{2}kx^2$.

$\dfrac{d}{dx}\left(\dfrac{1}{2}mv^2\right) = 0 = mg\sin37° - kx$, $\Rightarrow$ $x = \dfrac{mg\sin37°}{k} = \dfrac{(2.0 \text{ kg})(9.80 \text{ m/s}^2)\sin37°}{500 \text{ N/m}} = 0.024 \text{ m}$,

i.e., v is max $\boxed{\text{when the spring is compressed 0.024 m}}$.

Chapter 8 System of Particles

1. $X_{cm} = \dfrac{1}{M} \sum_i m_i x_i$

    $= \dfrac{(4.0 \text{ kg})(0) + (1.0 \text{ kg})(2.0 \text{ m})}{4.0 \text{ kg} + 1.0 \text{ kg}}$

    $= \boxed{0.40 \text{ m from the 4.0 kg mass}}$.

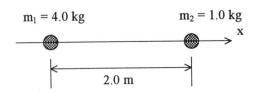

$m_1 = 4.0$ kg $m_2 = 1.0$ kg

2.0 m

7. Choose the free end of the rod as the origin ($x = 0$).

    The CM's of the rod and the ball are at L/2 and

    $(L + R)$, respectively. $X_{cm} = \dfrac{1}{M} \sum_i m_i x_i$

    $= \dfrac{(2M)(L/2) + M(L + R)}{2M + M} = \boxed{\dfrac{2L}{3} + \dfrac{R}{3}}$.

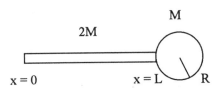

$x = 0$ $x = L$ R   2M   M

13. (a) The mass of the lid and the bottom is

    $m_1 = m_2 = \pi(15 \text{ cm})^2(10 \text{ g/cm}^2) = 7.07 \text{ kg}$,

    The mass of the side of the can is

    $m_3 = 2\pi(15 \text{ cm})(50 \text{ cm})(10 \text{ g/cm}^2) = 47.1 \text{ kg}$,

    The mass of the water is

    $m_4 = \pi(15 \text{ cm})^2(25 \text{ cm})(1.0 \text{ g/cm}^3) = 17.7 \text{ kg}$,

    $X_{cm} = \dfrac{1}{M} \sum_i m_i x_i$

    $= \dfrac{(7.07 \text{ kg})(50 \text{ cm}) + (7.07 \text{ kg})(0) + (47.1 \text{ kg})(25 \text{ cm}) + (17.7 \text{ kg})(12.5 \text{ cm})}{7.07 \text{ kg} + 7.07 \text{ kg} + 47.1 \text{ kg} + 17.7 \text{ kg}} = \boxed{22 \text{ cm from the bottom}}$,

    (b) $X_{cm} = \dfrac{(7.07 \text{ kg})(0) + (47.1 \text{ kg})(25 \text{ cm}) + (17.7 \text{ kg})(12.5 \text{ cm})}{7.07 \text{ kg} + 47.1 \text{ kg} + 17.7 \text{ kg}} = \boxed{19 \text{ cm from the bottom}}$.

    (c) Assume h cm of water needs to be added.

    The mass of extra water is $m_5 = \pi(15 \text{ cm})^2 h(1.0 \text{ g/cm}^3) = 0.707h \text{ kg}$,

    $X_{cm} = 22.2 \text{ cm} = \dfrac{(7.07 \text{ kg})(0) + (47.1 \text{ kg})(25 \text{ cm}) + (17.7 \text{ kg})(12.5 \text{ cm}) + (0.707h \text{ kg})(25 \text{ cm} + h/2)}{7.07 \text{ kg} + 47.1 \text{ kg} + 17.7 \text{ kg} + 0.707h \text{ kg}}$,

    Solve for $h = \boxed{21 \text{ cm}}$. So the can is filled to a height of 25 cm + 21 cm = 46 cm.

x

lid

m_1

m_3 side

m_4

m_2

bottom $x = 0$

19. Assume the surface mass density is σ.

    For the thin strip,

    $dm = \sigma y dx = \sigma A \cos x\, dx$,

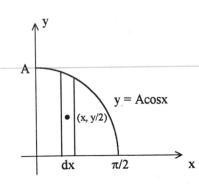

y

A

$y = A\cos x$

$(x, y/2)$

dx $\pi/2$ x

$$M = \int dm = \sigma A \int_0^{\pi/2} \cos x\, dx = \sigma A \sin x \Big|_0^{\pi/2} = \sigma A \,,$$

$$X_{cm} = \frac{1}{M} \int x\, dm = \frac{1}{\sigma A} \int_0^{\pi/2} x \sigma A \cos x\, dx = \int_0^{\pi/2} x \cos x\, dx = (x \sin x + \cos x)\Big|_0^{\pi/2} = \boxed{\frac{\pi}{2} - 1} \,,$$

$$Y_{cm} = \frac{1}{M} \int (y/2)\, dm = \frac{1}{\sigma A} \int_0^{\pi/2} \frac{A \cos x}{2} \cdot \sigma A \cos x\, dx = \frac{A}{2} \int_0^{\pi/2} \cos^2 x\, dx$$

$$= \frac{A}{2} \int_0^{\pi/2} \frac{1 + \cos 2x}{2}\, dx = \frac{A}{4}\left(x + \frac{\sin 2x}{2}\right)\Big|_0^{\pi/2} = \boxed{\frac{A\pi}{8}} \,.$$

25.　(a) $(\mathbf{r_0})_{cm} = \dfrac{1}{M} \sum_i m_i \mathbf{r}_i = \dfrac{(1.0\ \text{kg})[(2.0\ \mathbf{i} + 3.0\ \mathbf{j})\ \text{m}] + (3.0\ \text{kg})[(-6.0\ \mathbf{i} + 2.0\ \mathbf{j})\ \text{m}]}{1.0\ \text{kg} + 3.0\ \text{kg}} = \boxed{(-4.0\ \mathbf{i} + 2.3\ \mathbf{j})\ \text{m}} \,,$

(b) $(\mathbf{v_0})_{cm} = \dfrac{1}{M} \sum_i m_i \mathbf{v}_i = \dfrac{(1.0\ \text{kg})[(6.0\ \mathbf{i} - 4.0\ \mathbf{j})\ \text{m/s}] + (3.0\ \text{kg})[(-2.0\ \mathbf{i} + 1.0\ \mathbf{j})\ \text{m/s}]}{1.0\ \text{kg} + 3.0\ \text{kg}} = \boxed{-0.25\ \mathbf{j}\ \text{m/s}} \,,$

(c) $\mathbf{a}_{cm} = \dfrac{\mathbf{F}_{net}}{M} = \dfrac{\mathbf{F_1}}{M} = \dfrac{2.0\ \mathbf{j}\ \text{N}}{1.0\ \text{kg} + 3.0\ \text{kg}} = 0.50\ \mathbf{j}\ \text{m/s}^2 \,,$

$\mathbf{r}_{cm} = (\mathbf{r_0})_{cm} + (\mathbf{v_0})_{cm}t + \dfrac{1}{2}\mathbf{a}_{cm}t^2 = (-4.0\ \mathbf{i} + 2.25\ \mathbf{j})\ \text{m} + (-0.25\ \mathbf{j}\ \text{m/s})(5.0\ \text{s}) + \dfrac{1}{2}(0.50\ \mathbf{j}\ \text{m/s}^2)(5.0\ \text{s})^2$

$= \boxed{(-4.0\ \mathbf{i} + 7.3\ \mathbf{j})\ \text{m}} \,, \qquad \mathbf{v}_{cm} = (\mathbf{v_0})_{cm} + \mathbf{a}_{cm}t = -0.25\ \mathbf{j}\ \text{m/s} + (0.50\ \mathbf{j}\ \text{m/s}^2)(5.0\ \text{s}) = \boxed{2.3\ \mathbf{j}\ \text{m/s}} \,,$

(d) $\boxed{\text{No change}}$ since they are an internal forces.

31.　Assume the surface mass density is σ.

(a) x axis: $dI_x = (dm)y^2 = \sigma \cdot (b - x)dy\, y^2 = \sigma(b - x)y^2 dy$

$= \sigma\left(b - \dfrac{by}{a}\right)y^2 dy,$

$I_x = \int_0^a \sigma\left(b - \dfrac{by}{a}\right)y^2 dy = \left[\dfrac{\sigma b}{3}y^3 - \dfrac{\sigma b}{4a}y^4\right]\Big|_0^a$

$= \dfrac{1}{12}\sigma b a^3 = \dfrac{1}{6}\left(\dfrac{1}{2}\sigma b a\right)a^2 = \boxed{\dfrac{1}{6}Ma^2} \,.$

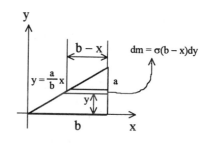

(b) y axis: $\qquad dI_y = x^2 \sigma y\, dx = x^2 \sigma \dfrac{a}{b} x\, dx,$

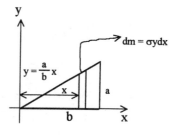

$$I_y = \int_0^b \sigma \frac{a}{b} x^3 \, dx = \frac{1}{4} \sigma ab^3 = \frac{1}{2} \left(\frac{1}{2} \sigma ab \right) b^2 = \boxed{\frac{1}{2} Mb^2}.$$

37. $I_{cm} = \frac{2}{3} MR^2$,

Spherical shell

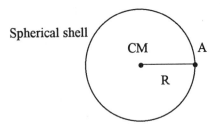

$$I_A = I_{cm} + Mh^2 = \frac{2}{3} MR^2 + MR^2 = \boxed{\frac{5}{3} MR^2} .$$

43.   The center of mass follows the projectile trajectory since the only external force is the gravitational force. The piece with zero initial velocity (relative to the explosion) lands at R/2 from where it is launched, where R is the range of the projectile. The center of mass should land at R . Since the two pieces have equal mass, the second piece will land at $\frac{3R}{2}$ from where it is launched because,

$$X_{cm} = R = \frac{1}{M} \sum_i m_i x_i = \frac{(M/2)(R/2) + (M/2)(3R/2)}{M}, \qquad R = \frac{v_0^2 \sin 2\theta}{g} = \frac{(500 \text{ m/s})^2 \sin 74^\circ}{9.80 \text{ m/s}^2} = 24.5 \text{ km} ,$$

Therefore the second piece lands at $\frac{3(24.5 \text{ km})}{2} = \boxed{37 \text{ km from where it is launched}}$.

Chapter 9 Rotational Mechanics I

1.   (a) $\omega = \frac{\Delta\theta}{\Delta t} = \frac{2\pi \text{ rad}}{1 \text{ day}} = \frac{2\pi \text{ rad}}{86400 \text{ s}} = \boxed{7.27 \times 10^{-5} \text{ rad/s}}$.

(b) $\omega = \frac{\Delta\theta}{\Delta t} = \frac{2\pi \text{ rad}}{1 \text{ year}} = \frac{2\pi \text{ rad}}{(365)(86400 \text{ s})} = \boxed{1.99 \times 10^{-7} \text{ rad/s}}$.

7. $\theta - \theta_0 = \omega_o t + \frac{1}{2}\alpha t^2$, $\qquad \Rightarrow \qquad 500 \text{ rad} = \omega_o (6.0 \text{ s}) + \frac{1}{2} \alpha (6.0 \text{ s})^2$,

$\omega = \omega_o + \alpha t$, $\qquad \Rightarrow \qquad 100 \text{ rad/s} = \omega_o + \alpha(6.0 \text{ s})$,

Solve for   (a) $\alpha = \boxed{5.6 \text{ rad/s}^2}$,   and   (b) $\omega_o = \boxed{67 \text{ rad/s}}$.

13.   (a) $\omega = \omega_0 + \alpha t = 2.0 \text{ rad/s} + (1.0 \text{ rad/s}^2)(5.0 \text{ s}) = \boxed{7.0 \text{ rad/s}}$,

(b) $\theta - \theta_0 = \omega_0 t + \frac{1}{2}\alpha t^2 = (2.0 \text{ rad/s})(5.0 \text{ s}) + \frac{1}{2}(1.0 \text{ rad/s}^2)(5.0 \text{ s})^2 = \boxed{23 \text{ rad}}$,

(c) $a_\theta = \alpha r = (1.0 \text{ rad/s}^2)(0.10 \text{ m}) = \boxed{0.10 \text{ m/s}^2}$, $a_r = -r\omega^2 = -(0.10 \text{ m})(7.0 \text{ rad/s})^2 = \boxed{-4.9 \text{ m/s}^2}$.

(d) $a_\theta = \boxed{0.10 \text{ m/s}^2}$, $\qquad \omega = \omega_0 + \alpha t = 2.0 \text{ rad/s} + (1.0 \text{ rad/s}^2)(0.50 \text{ s}) = 2.5 \text{ rad/s}$,

So, $\qquad a_r = -r\omega^2 = -(0.10 \text{ m})(2.5 \text{ rad/s})^2 = \boxed{-0.63 \text{ m/s}^2}$.

19. (a) $\Sigma\tau = \Sigma r_\perp F = [(1.0 \text{ m})\cos 60°](20 \text{ N}) + (4.0 \text{ m})(40 \text{ N}) - (3.0 \text{ m})(20 \text{ N}) + [(5.0 \text{ m})\sin 53°](30 \text{ N})$

$= \boxed{230 \text{ N·m}}$,

(b) $\Sigma\tau = \mathbf{r} \times \mathbf{F} = (1.0 \, \mathbf{j} \text{ m}) \times (20 \text{ N})[(-\cos 60° \, \mathbf{i} + \sin 60° \, \mathbf{j})] + (4.0 \, \mathbf{i} \text{ m}) \times (40 \, \mathbf{j} \text{ N}) + (-3.0 \, \mathbf{j} \text{ m}) \times (-20 \, \mathbf{i} \text{ N})$

$+ (-5.0 \, \mathbf{i} \text{ m}) \times [(30 \text{ N})(-\cos 53° \, \mathbf{i} - \sin 53° \, \mathbf{j})] = 10 \, \mathbf{k} \text{ N·m} + 160 \, \mathbf{k} \text{ N·m} - 60 \, \mathbf{k} \text{ N·m} + 120 \, \mathbf{k} \text{ N·m}$

$= \boxed{230 \, \mathbf{k} \text{ N·m}}$.

25. $\omega_0 = 500 \text{ rev/min} = 52.4 \text{ rad/s}$, $\qquad \alpha = \dfrac{\Delta\omega}{\Delta t} = \dfrac{0 - 52.4 \text{ rad/s}}{2.0(60 \text{ s})} = -0.436 \text{ rad/s}^2$,

$\Sigma\tau_f = I\alpha = (100 \text{ kg·m}^2)(-0.436 \text{ rad/s}^2) = \boxed{-44 \text{ N·m}}$.

31. $\Sigma\tau = -f_k R = -\mu_k N R = -0.50(10 \text{ N})(0.20 \text{ m}) = -1.0 \text{ N·m} = I\alpha$,

$I = \dfrac{1}{2}MR^2 = \dfrac{1}{2}(10 \text{ kg})(0.20 \text{ m})^2 = 0.20 \text{ kg·m}^2$, $\qquad$ So, $\qquad \alpha = \dfrac{-1.0 \text{ N·m}}{0.20 \text{ kg·m}^2} = -5.0 \text{ rad/s}^2$,

$\omega = \omega_0 + \alpha t$, $\qquad \Rightarrow \qquad 0 = 100 \text{ rad/s} + (-5.0 \text{ rad/s}^2)t$, $\qquad t = \boxed{20 \text{ s}}$.

37. Minimum work means minimum moment of inertia since $W = \Delta T = \dfrac{1}{2}I(\omega^2 - 0)$,

For the rod, $\qquad I_{cm} = \dfrac{1}{12}ML^2$, $\qquad I_a = I_{cm} + Mh^2$, $\qquad I = m_1 d^2 + m_2(L - d)^2 + \dfrac{1}{12}ML^2 + M(\dfrac{L}{2} - d)^2$,

$\dfrac{dI}{dd} = 2m_1 d - 2m_2(L - d) - 2M(\dfrac{L}{2} - d) = 0$, $\qquad d = \boxed{\dfrac{M + 2m_2}{m_1 + m_2 + M} \cdot \dfrac{L}{2}}$.

43. Since the belt is not slipping, $\qquad a_1 = \alpha_1 r_1 = a_2 = \alpha_2 r_2$, $\qquad \Rightarrow \qquad \alpha_1 = \dfrac{0.30 \text{ m}}{0.20 \text{ m}}\alpha_2 = 1.5\alpha_2$,

For the small wheel, $\qquad \Sigma\tau = 10 \text{ N·m} + (T' - T)r_1 = I_1\alpha_1$, or, $\qquad 50 \text{ N} + T' - T = (20 \text{ kg·m})\alpha_1$,

For the large wheel,

$\Sigma\tau = (T - T')r_2 = I_2\alpha_2$,

or, $\qquad T - T' = (66.7 \text{ kg·m})\alpha_2$,

Solve for $\alpha_2 = \boxed{0.52 \text{ rad/s}^2}$,

and $\qquad \alpha_1 = \boxed{0.78 \text{ rad/s}^2}$.

49. (a) Use $v = \omega r$, $\qquad v_b = \omega R_e$,

$v_t = \omega(R_e + h)$, $\qquad$ So, $\qquad v_t - v_b = \boxed{\omega h}$,

(b) The relative horizontal velocity is $v_t - v_b$,

In the vertical direction, $x - x_0 = v_0 t + \frac{1}{2}at^2$, $\Rightarrow$ $h = 0 + \frac{1}{2}gt^2$, $t = \sqrt{\dfrac{2h}{g}}$,

So, $d = (v_t - v_b)t = \boxed{\omega h \sqrt{\dfrac{2h}{g}}}$,

(c) $\omega = \dfrac{2\pi \text{ rad}}{1 \text{ day}} = \dfrac{2\pi \text{ rad}}{86400 \text{ s}} = 7.27 \times 10^{-5}$ rad/s,

So, $d = (7.27 \times 10^{-5} \text{ rad/s})(30 \text{ m}) \sqrt{\dfrac{2(30 \text{ m})}{9.80 \text{ m/s}^2}} = \boxed{5.4 \text{ mm}}$.

Chapter 10 Rotational Mechanics II

1. $v_{cm} = \omega R$, $\Rightarrow$ $\omega = \dfrac{v_{cm}}{R} = \dfrac{(90 \text{ km/hr})(1000 \text{ m/km})(1 \text{ hr/3600 s})}{0.375 \text{ m}} = \boxed{67 \text{ rad/s}}$.

7. $\Sigma F_y = N - Mg = M(a_{cm})_y = 0$, $\quad N = Mg$,

$\Sigma F_x = T - f_k = T - \mu_k N = T - \mu_k Mg = M(a_{cm})_x$,

$(a_{cm})_x = \boxed{\dfrac{T}{M} - \mu_k g}$,

$\Sigma\tau = TR - f_k R = (T - \mu_k Mg)R = I_{cm}\alpha = \frac{1}{2}MR^2\alpha$,

$\alpha = \boxed{\dfrac{2}{R}(\dfrac{T}{M} - \mu_k g)}$.

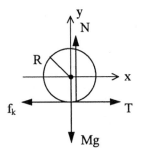

13. (a) $W = \Delta T = 0 - (\frac{1}{2}mv^2_{cm} + \frac{1}{2}I_{cm}\omega^2) = -\frac{1}{2}(6.0 \text{ kg})(10 \text{ m/s})^2 - \frac{1}{2}[(6.0 \text{ kg})R^2] \cdot (\dfrac{10 \text{ m/s}}{R})^2 = \boxed{-600 \text{ J}}$,

(b) Choose the CM when at the bottom as $U = 0$ and use

$T + U = $ constant.

$600 \text{ J} + 0 = 0 + (6.0 \text{ kg})(9.80 \text{ m/s}^2)h$,

Solve for $\quad h = 10.2 \text{ m}$,

So, $\quad d = \dfrac{h}{\sin 30°} = \dfrac{10.2 \text{ m}}{\sin 30°} = \boxed{20 \text{ m}}$.

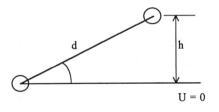

19. $\Sigma F_y = N + F\sin 37° - Mg = M(a_{cm})_y = 0$,

$N = Mg - F\sin 37°$,

$\Sigma F_x = F\cos 37° - f_s = M(a_{cm})_x$,

$\Sigma\tau = f_s R = I_{cm}\alpha = (\frac{1}{2}MR^2) \cdot \dfrac{(a_{cm})_x}{R}$,

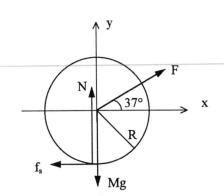

$$F = \frac{3f_s}{\cos 37°} \le \frac{3}{\cos 37°} \mu_s N = \frac{3(0.40)}{\cos 37°} (Mg - F\sin 37°) ,$$

Therefore, $\quad F \le \dfrac{1.50Mg}{1.90} = \boxed{0.79Mg}$.

25. Assume the roll accelerates forward relative to the ground with an acceleration of a' , then it accelerates backwards relative to the truck with an acceleration of (a − a').

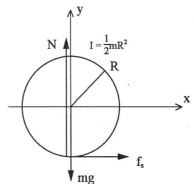

Also $\alpha R = a - a'$, and $I = \frac{1}{2}mR^2$, $\Sigma F_x = f_s = ma'$,

$$\Sigma\tau = f_s R = I\alpha = I\frac{a - a'}{R} , \text{ or } f_s = \frac{I}{R^2}(a - a') = \frac{m(a - a')}{2} ,$$

Solve for $\quad a' = \dfrac{a}{3}$,

Now use $x - x_0 = v_0 t + \frac{1}{2}at^2$, $\quad$ Relative to the truck,

$$d = 0 + \frac{1}{2}(a - a')t^2 = \frac{1}{3}at^2 , \qquad t = \sqrt{\frac{3d}{a}} ,$$

Therefore, $\quad S = 0 + \dfrac{1}{2}at^2 = \dfrac{1}{2}a\left(\dfrac{3d}{a}\right) = \boxed{1.5d}$.

31. (a) $\quad \Sigma F_x = -f_s = Ma$,

$$\Sigma\tau = f_s R - \tau = I\alpha = \frac{1}{2}mR^2 \frac{a}{R} ,$$

or, $\quad f_s - \dfrac{\tau}{R} = \dfrac{1}{2}ma$,

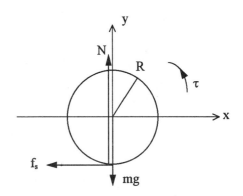

Solve for $\quad a = \boxed{-\dfrac{2\tau}{3mR}}$,

(b) $\Sigma F_y = N - mg = ma_y = 0$, $\qquad N = mg$,

$$\Sigma\tau = f_s R - \tau = \frac{1}{2}mR^2 \cdot \left(-\frac{2\tau}{3mR^2}\right) = -\frac{\tau}{3} ,$$

So, $\quad \tau = \dfrac{3R}{2}f_s \le \dfrac{3R}{2}\mu_s N = \dfrac{3\mu_s mgR}{2}$,

Therefore, $\quad \tau_{max} = \boxed{\dfrac{3\mu_s mgR}{2}}$, $\qquad |a_{max}| =$

$$\frac{2}{3mR} \cdot \frac{3\mu_s mgR}{2} = \boxed{\mu_s g} .$$

(c) If $\tau > \tau_{max}$, kinetic friction replaces static friction.

$\Sigma F_x = -f_k = -\mu_k N = -\mu_k mg = ma$,

$a = \boxed{-\mu_k g}$,

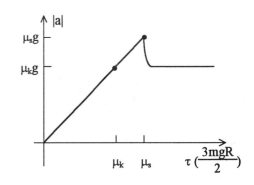

(d) $a = -\mu_k g$ is also for minimum τ ,

$\dfrac{2\tau_{min}}{3mR} = \mu_k g$, $\quad$ So, $\quad \tau_{min} = \dfrac{3\mu_k mgR}{2}$,

Therefore, $\quad \boxed{\dfrac{3\mu_k mgR}{2} < \tau < \dfrac{3\mu_s mgR}{2}}$,

(e) See graph on previous page.

(f) Just enough pressure on the brake paddle so the wheels do not lock and slide.

Chapter 11 Equilibrium of a Rigid Body

1. The second condition of equilibrium: $\Sigma\tau = 0$.

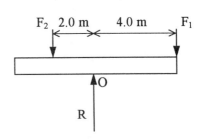

Use "O" as axis of rotation. $-(2.0)F_2 + (4.0)F_1 = 0$,

$F_1 = m_1 g = 40g$ N, $F_2 = m_2 g$ N,

R = reaction of support,

Thus, $-(2.0)m_2 g + (4.0)(40g) = 0$,

or, $m_2 = \dfrac{(4.0)(40)g}{(2.0)g} = 80$ kg.

7. Free-body diagram of chain:

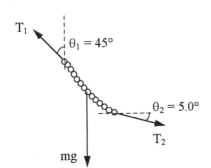

The first condition of equilibrium: $\Sigma\mathbf{F} = 0$,

or, $\Sigma F_x = 0$ and $\Sigma F_y = 0$.

$\Sigma F_x = T_2\cos\theta_2 - T_1\sin\theta_1 = 0$,

$\Sigma F_y = T_1\cos\theta_1 - T_2\sin\theta_2 - mg = 0$,

From ΣF_x, $T_2 = T_1\left[\dfrac{\sin\theta_1}{\cos\theta_2}\right] = T_1\left[\dfrac{\sin45°}{\cos5°}\right] = 0.71T_1$,

From ΣF_y, $m = \dfrac{T_1}{g}[\cos\theta_1 - 0.71\sin\theta_2]$

$= T_1\left[\dfrac{0.701 - (0.71)(0.087)}{9.8}\right] = 0.066T_1$,

Free-body diagram of m_1: $T_1 = m_1 g = (0.10)(9.8) = 0.98$ N.

Free-body diagram of m_2: $T_2 = m_2 g$.

Finally, $m = 0.066T_1 = (0.066)(0.98) = 0.065$ kg,

and $T_2 = 0.71T_1 = (0.71)(0.98) = 0.70$ N,

and $m_2 = \dfrac{T_2}{g} = \dfrac{0.70}{9.8} = 0.071$ kg.

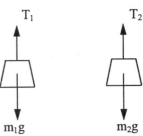

13. $\Sigma\tau = (2000)(2.0) + (3000)(2.0) - (5000)(1.0)$

$= 5000$ N·m $\neq 0$, so not possible.

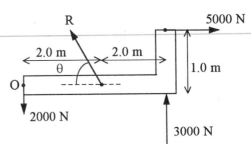

19.   Free-body diagram of wheel:

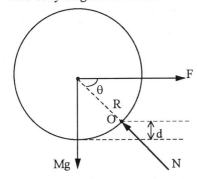 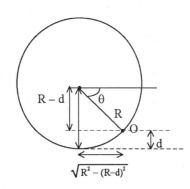

The second condition of equilibrium: $\Sigma\tau = 0$ (using point O as axis of rotation).

$\Sigma\tau = (Mg)(R\cos\theta) - (F)(R\sin\theta) = 0,$ or, $F = Mg\left(\dfrac{\cos\theta}{\sin\theta}\right) = mg$

But $\dfrac{\sin\theta}{\cos\theta} = \tan\theta = \dfrac{R - d}{\sqrt{R^2 - (R-d)^2}}$.   Thus, $m = \dfrac{M\sqrt{R^2 - (R-d)^2}}{R - d} = \dfrac{M\sqrt{2Rd - d^2}}{R - d}$.

25.   The first condition of equilibrium:

$\Sigma F_x = T\cos 20° - H = 0,$

$\Sigma F_y = V - 300 - T\sin 20° = 0.$

The second condition of equilibrium:

$\Sigma\tau = 0,$ using hinge as axis of rotation.

$\Sigma\tau = (300)(l/2)\cos 30° - (T\sin 10°)(l/2) = 0,$

Thus, $T = \dfrac{(300)\cos 30°}{\sin 10°} = 1500$ N.

and $H = T\cos 20° = (1500)\cos 20° = 1410$ N,

$V = 300 + T\sin 20° = 300 + (1500)\sin 20° = 813$ N.   The resultant force on trap door at hinge:

$R = \sqrt{V^2 + H^2} = \sqrt{(813)^2 + (1410)^2} = 1630$ N, $\theta = \arctan\left(\dfrac{V}{H}\right) = \arctan\left(\dfrac{813}{1410}\right) = 30°.$

31.   The second condition of equilibrium:

$\Sigma\tau = 0.$ Using point "A" as axis of rotation.

$\Sigma\tau = Nx - (mg\cos\theta)(w/2) + (mg\sin\theta)(h/2) = 0,$

The first condition of equilibrium: $\Sigma F_x = f - mg\sin\theta = 0,$

$\Sigma F_y = N - mg\cos\theta = 0.$

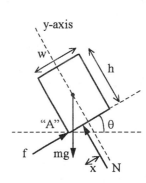

(a) Thus, $x = \dfrac{mg}{2}\dfrac{(w\cos\theta - h\sin\theta)}{mg\cos\theta} = \dfrac{1}{2}(w - h\tan\theta).$

(b) Maximum value of angle θ occurs when $x = 0,$

or, $w = h\tan\theta,$ $\theta = \arctan\left(\dfrac{w}{h}\right).$

37. Since $Y = \dfrac{FL}{A\Delta L}$, then $A = \dfrac{FL}{Y\Delta L} = \dfrac{(2.5\times10^4)(9.8)(25)}{(2.0\times10^{11})(10^{-2})} = 3.1\times10^{-3}$ m².

Since $A = \dfrac{\pi D^2}{4}$, $D = \sqrt{\dfrac{4A}{\pi}} = \sqrt{\dfrac{4(3.1\times10^{-3})}{\pi}} = 6.24\times10^{-2}$ m = 6.2 cm.

43. Since volume $V = a^3$, where a is the length of a side, the change in volume, $dV = 3a^2 da$.

The fraction change in volume, $\dfrac{dV}{V} = \dfrac{3a^2 da}{a^3} = \dfrac{3}{a} da$,

or, $\dfrac{dV}{V} = -\dfrac{(3)(0.005)}{(50)} = -3\times10^{-4}$, where da = 49.995 − 50.000 = −0.005.

Also, $\dfrac{\Delta F}{A} = \dfrac{7.0\times10^6}{(0.5)^2} = 2.8\times10^7$ N/m², Thus, bulk modulus, $B = -\dfrac{(2.8\times10^7)}{(-3\times10^{-4})} = 9.3\times10^{10}$ N/m².

49. The first condition of equilibrium:

$\sum F_x = f - Mg\sin\theta + F\cos\theta = 0$, (1),

$\sum F_y = N - Mg\cos\theta - F\sin\theta = 0$, (2).

The second condition of equilibrium, using "O" as axis:

$\sum \tau = Mg(R\sin\theta) - (F\cos\theta)(R) = 0$, (3).

From (3): $F = Mg\tan\theta$,

From (1): $f = Mg\sin\theta - F\cos\theta = Mg\sin\theta - Mg\sin\theta = 0$.

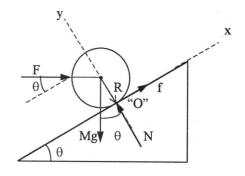

55. Free-body diagram of one orderly:

(a) The second condition of equilibrium:

$\sum \tau_P = T(11) - (400)(13) - (350)(29) = 0$,

T = 1395 N.

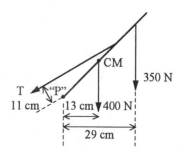

(b) Free-body diagram with foreheads touching (no back tension):

$\sum \tau_P = F(48) - (400)(13) - (350)(29) = 0$,

F = 320 N.

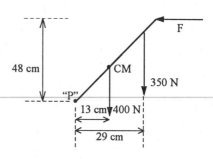

(c) Free-body diagram with foreheads touching (reduced back tension):

$\sum \tau_P = T(11) + (200)(48) - (400)(13) - (350)(29) = 0,$

$T = 523$ N.

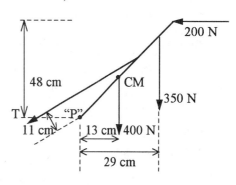

Chapter 12 Momentum

1. $p_{16} = \left(\dfrac{w}{g}\right)v = 16\left(\dfrac{v}{g}\right),$ and $p_{10} = 10\left(\dfrac{v}{g}\right),$ The ratio: $\dfrac{p_{16}}{p_{10}} = \dfrac{16}{10} = 1.6.$

7. $p_{before} = p_{after},$ $0 = (0.5)(20) + (80)v,$ $v = -0.13$ m/s (relative to capsule)

    Astronaut moves in a direction opposite to that of the hammer.

13. $\mathbf{p_{before}} = \mathbf{p_{after}},$ $(90)(-8.0\ \mathbf{i}) + (120)(6.5)\left(\dfrac{\mathbf{i} + \mathbf{j}}{\sqrt{2}}\right) = (90 + 120)\mathbf{v},$ $\mathbf{v} = (-0.80\ \mathbf{i} + 2.6\ \mathbf{j})$ m/s.

    $K_{initial} = \dfrac{1}{2}(90)(8.0)^2 + \dfrac{1}{2}(120)(6.5)^2 = 5400$ J, $K_{final} = \dfrac{1}{2}(90 + 120)(0.8^2 + 2.6^2) = 780$ J,

    $\Delta K = 5400 - 780 = 4620$ J lost.

19. (a) Since $V_{CM} = 0$ due to the lack of external forces,

    X_{CM} is fixed at $X_{CM} = \dfrac{m_1(0) + m_2 d}{m_1 + m_2} = \left(\dfrac{m_2}{m_1 + m_2}\right)d.$

    (b) Still $V_{CM} = 0$, so they will meet at the same CM place.

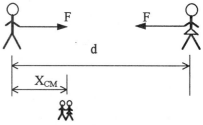

25. Initial momentum,

    $\mathbf{p_1} = (0.05)(40)(\cos 30°\ \mathbf{i} - \sin 30°\ \mathbf{j}) = (1.73\ \mathbf{i} - 10.0\ \mathbf{j})$ N·s.

    $\mathbf{p_2} = (0.05)(40)(-\cos 30°\ \mathbf{i} - \sin 30°\ \mathbf{j}) = (-1.73\ \mathbf{i} - 10.0\ \mathbf{j})$ N·s.

    (a) Change in momentum $\Delta \mathbf{p} = \mathbf{p_2} - \mathbf{p_1}$

    $= (-1.73\ \mathbf{i} - 10.0\ \mathbf{j}) - (1.73\ \mathbf{i} - 10.0\ \mathbf{j}) = -3.46\ \mathbf{i}$ N·s.

    (b) Impulse on wall $I_{wall} = -I_{ball} = -\Delta \mathbf{p}_{ball} = 3.46\ \mathbf{i}$ N·s.

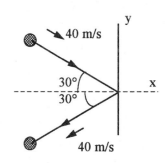

31.   Conservation of momentum: $p_{before} = p_{after}$. $m_p v_p = m_p v_p' + m_{He} v_{He}'$.

Since $m_{He} = 4m$,   and $m_p = m$, $v_p = v_p' + 4 v_{He}'$,   (1).

Conservation of kinetic energy for elastic collision: $\frac{1}{2} m_p v_p^2 = \frac{1}{2} m_p (v_p')^2 + \frac{1}{2} m_{He} (v_{He}')^2$.

or, $v_p^2 = (v_p')^2 + 4(v_{He}')^2$,   (2).

From (1): $v_{He}' = \dfrac{v_p - v_p'}{4}$,   (2) becomes $v_p^2 = (v_p')^2 + \dfrac{1}{4}(v_p - v_p')^2$,

Or, $5(v_p')^2 - (2v_p) v_p' - 3 v_p^2 = 0$,   Use quadratic formula $v_p' = v_p$; $-0.6 v_p$.

Since proton rebounds, $v_p' = -0.6 v_p$,   or, $v_p' = -(0.6)(6.0 \times 10^6) = -3.6 \times 10^6$ m/s,

and $v_{He}' = \dfrac{1}{4}(v_p - v_p') = \dfrac{1}{4}(6.0 + 3.6) \times 10^6 = 2.4 \times 10^6$ m/s.

37.   From Example 12-11, $v = u \ln \dfrac{m_o}{m_o - rt} - gt = (4000) \ln \dfrac{4000}{4000 - (10)(120)} - 9.8(120) = 250$ m/s.

43.   Conservation of momentum: $(3m)(5.0 \times 10^3 \mathbf{i}) = m\mathbf{v_1} + m\mathbf{v_2} + m\mathbf{v_3}$,

where $\mathbf{v_1} = \dfrac{\mathbf{r_1}}{3} = \dfrac{20\,\mathbf{i} - 10\,\mathbf{j}}{3} \times 10^3$ m/s, $\mathbf{v_2} = \dfrac{\mathbf{r_2}}{3} = \dfrac{-30\,\mathbf{i} + 25\,\mathbf{k}}{3} \times 10^3$ m/s,

Thus, $\mathbf{v_3} = (15.0 \times 10^3)\mathbf{i} - \mathbf{v_1} - \mathbf{v_2} = (15.0 \times 10^3)\mathbf{i} - \dfrac{20\,\mathbf{i} - 10\,\mathbf{j}}{3} \times 10^3 - \dfrac{-30\,\mathbf{i} + 25\,\mathbf{k}}{3} \times 10^3$

$= (18.3\,\mathbf{i} + 3.3\,\mathbf{j} - 8.3\,\mathbf{k}) \times 10^3$ m/s,   and $\mathbf{r_3} = \mathbf{v_3} t = 3\mathbf{v_3} = (55.0\,\mathbf{i} + 10.0\,\mathbf{j} - 25.0\,\mathbf{k})$ km.

49.   Conservation of momentum: (ignore rotation) $m_1 v_1 = m_1 u_1 + m_2 u_2$,   (1)

For elastic collision, velocity of approach before collision = velocity of separation after collision.

Thus, $v_1 = u_2 - u_1$,   (2).

Substituting: $u_2 = v_1 + u_1$ into (1)

$m_1 v_1 = m_1 u_1 + m_2 (v_1 + u_1)$,

or, $u_1 = \dfrac{m_1 - m_2}{m_1 + m_2} v_1$,

and $u_2 = \dfrac{2 m_1}{m_1 + m_2} v_1$.

(a) $K_2 = \dfrac{1}{2} m_2 u_2^2 = \dfrac{1}{2} m_2 \left(\dfrac{2 m_1}{m_1 + m_2}\right)^2 v_1^2 = \dfrac{2 m_1^2 m_2}{(m_1 + m_2)^2} v_1^2$.

(b) Let $\dfrac{m_1}{m_2} = r$. Then $K_2 = 2 m_1 v_1^2 \left[\dfrac{r}{(1 + r)^2}\right]$.

To find maximum: $\dfrac{d}{dr}\left[\dfrac{r}{(1 + r)^2}\right] = \left[\dfrac{1}{(1 + r)^2} - \dfrac{2r}{(1 + r)^3}\right] = 0$.   or, $r = 1$.

Thus, kinetic energy of m_2 is maximum when $m_2 = m_1$.

Chapter 13 Angular Momentum

1. $l = \mathbf{r} \times \mathbf{p} = m\mathbf{r} \times \mathbf{v} = (0.20)(2)(5)\,(\mathbf{j} \times \mathbf{i}) = (-2.0\,\mathbf{k})\ \text{kg·m}^2/\text{s}.$

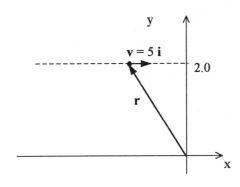

7.   Since $\mathbf{v}$ is a constant, $r_\perp$ is a constant, so $mr_\perp v$ is a constant.

13. $l = mvr = mr^2\omega = (5.98\times10^{24})(1.5\times10^{11})^2\ \dfrac{2\pi}{365\times24\times3600} = 2.68\times10^{40}\ \text{kg·m}^2/\text{s},$

   So, $\dfrac{l_{\text{orbit}}}{L_{\text{rot}}} = \dfrac{2.68\times10^{40}}{7.1\times10^{33}} = 3.8\times10^{6}$

19.   Angular Impulse, $\mathbf{A} = \Delta\mathbf{L}$,   so, $\Delta\mathbf{L} = 2.0\,\mathbf{j}\ \text{kg·m}^2/\text{s}$,

   $\mathbf{L}_{\text{final}} - \mathbf{L}_{\text{initial}} = \Delta\mathbf{L}$, $\Rightarrow$ $0.80\,\mathbf{j} - \mathbf{L_i} = 2.0\,\mathbf{j}$,   So, $\mathbf{L_1} = -1.2\,\mathbf{j}\ \text{kg·m}^2/\text{s}.$

25.   (a) $I_1\omega_1 = I_2\omega_2$, $\omega_1 = 1.0\ \text{rev/s} = 2\pi\ \text{rad/s} = 6.28\ \text{rad/s}$

   $(5.0)(1.0\,\dfrac{\text{rev}}{\text{s}}) = (1.5)\,\omega_2$, $\Rightarrow$ $\omega_2 = \dfrac{5.0}{1.5} = 3\tfrac{1}{3}\ \text{rev/s} = 20.9\ \text{rad/s}.$

   (b) $\dfrac{I_2\,\omega^2_2/2}{I_1\,\omega^2_1/2} = \dfrac{1.5(20.9)^2}{5(6.28)^2} = 3.3.$ So the kinetic energy increases by a factor of 3.3.

31.   Conservation of Angular Momentum. $I_{\text{Disk}} = \tfrac{1}{2}MR^2 = \tfrac{1}{2}(0.10)(0.10)^2 = 5.0\times10^{-4}\ \text{kg·m}^2$

   $I_{\text{Bug}} = mR^2 = (0.020)(0.10)^2 = 2.0\times10^{-4}\ \text{kg·m}^2.$

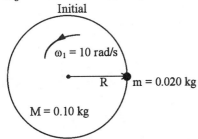

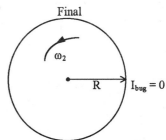

(a) $(I_{disk} + I_{Bug}) \omega_1 = I_{Disk} \omega_2$, $\qquad \omega_2 = \dfrac{5.0 \times 10^{-4} + 2.0 \times 10^4}{5.0 \times 10^{-4}}(10) = 14.0$ rad/s.

(b) $\Delta T = T_{final} - T_{initial} = \dfrac{1}{2}(5.0 \times 10^{-4})(14.0)^2 - \dfrac{1}{2}(5.0 + 2.0)(10^{-4})(10)^2 = 0.014$ J.

(c) $\omega_3 = \dfrac{(14.0)(5.0 \times 10^{-4})}{(5.0 + 2.0)(10^{-4})} = 10$ rad/s. $\qquad$ (d) $T_3 = \dfrac{1}{2}(I_{Disk} + I_{Bug})\omega_3{}^2 = 0.035$ J.

(e) Work of the bug's feet on the disk.

37. $\quad$ Conservation of Angular Momentum: $\qquad mLv = (I_{Bug} + I_{stick})\omega = \left[mL^2 + \dfrac{1}{3}(10m)L^2 \right]\omega$,

So, $\qquad \omega = \dfrac{mLv}{(m + 10m/3)L^2} = \dfrac{3v}{13L}$.

After collision, conservation of mechanical energy:

$\Delta T = -\dfrac{1}{2}(I_{Bug} + I_{stick})\omega^2$,

$\Delta U = g\left[10m(L/2) + mL \right](1 - \cos 5°)$,

So, $\qquad \dfrac{1}{2}\left(mL^2 + \dfrac{1}{3}(10m)L^2 \right)\left[\dfrac{3v}{13L} \right]^2$

$\qquad = g\left[10m(L/2) + mL \right](1 - \cos 5°)$,

Solving for $\qquad L = \dfrac{(1/2)(m + 10m/3)(3)^2 v^2}{(g)(m + 10m/2)(13)^2(1 - \cos 5°)}$

$\qquad = 0.52$ m.

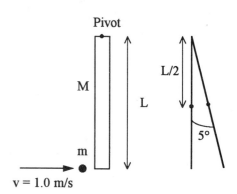

Pivot

M

m

$v = 1.0$ m/s

L

L/2

5°

Chapter 14 Oscillatory Motion

1. $\quad$ Force constant of Spring: $\qquad k = \dfrac{F}{x} = \dfrac{(150)(9.8)}{0.020} = 7.35 \times 10^4$ N/m.

$\quad$ Period, $\qquad T = 2\pi\sqrt{\dfrac{m}{k}} = 2\pi\sqrt{\dfrac{150 + 1200}{7.35 \times 10^4}} = 0.85$ s.

7. $\quad$ Angular frequency, $\qquad \omega = \dfrac{2\pi}{T} = \dfrac{2\pi}{0.25} = 8\pi$ rad/s.

$\quad$ Maximum velocity, $\qquad v_{max} = A\omega = (0.10)(8\pi) = 2.51$ m/s.

$\quad$ Total Energy = Maximum kinetic energy $= \dfrac{1}{2}m(v_{max})^2 = \dfrac{1}{2}(2.0)(2.51)^2 = 6.32$ J.

13. $\quad$ Let $\quad M = 0.150$ kg, $\quad A = 0.05$ m, $\quad T = 0.25$ s. $\quad$ Then, $\quad \omega = \dfrac{2\pi}{T} = \dfrac{2\pi}{0.25} = 8\pi$ rad/s.

(a) $v_{max} = A\omega = (0.05)(8\pi) = 1.26$ m/s, $\qquad a_{max} = A\omega^2 = (0.05)(8\pi)^2 = 31.6$ m/s^2.

(b) Since $x(t) = A\sin(\omega t)$, $\qquad 3.0 = 5.0\sin(\omega t)$, $\qquad$ or, $\quad (\omega t) = \arcsin(0.6) = 0.644$ rad.

Then, $v(t) = \dfrac{dx}{dt} = A\omega\cos(\omega t)$, and $a(t) = \dfrac{dv}{dt} = -A\omega^2\sin(\omega t) = -\omega^2 x$.

Thus, at $x = 3.0$ cm: $v = (0.05)(8\pi)\cos(0.644) = 1.00$ m/s, and $a = -(8\pi)^2(0.03) = -18.9$ m/s^2.

(c) Since $T = 2\pi\sqrt{\dfrac{m}{k}}$, $\quad k = m\left(\dfrac{2\pi}{T}\right)^2 = (0.150)\left(\dfrac{2\pi}{0.25}\right)^2 = 94.7$ N/m.

(d) $E = T_{max} = U_{max} = \dfrac{1}{2}m(v_{max})^2 = \dfrac{1}{2}(0.150)(1.26)^2 = 0.119$ J.

19. Period, $T = 2\pi\sqrt{\dfrac{L}{g}} = 2\pi\sqrt{\dfrac{2.0}{9.8}} = 2.84$ s. Angular frequency, $\omega = \dfrac{2\pi}{T} = 2.21$ rad/s.

(a) Maximum angular velocity, $\left(\dfrac{d\theta}{dt}\right)_{max} = A\omega = (10°)\left(\dfrac{\pi}{180°}\right)(2.21) = 0.39$ rad/s.

(b) Maximum linear speed, $v_{max} = (l)\left(\dfrac{d\theta}{dt}\right)_{max} = (2.0)(0.39) = 0.77$ m/s.

(c) Total energy $= U_{max} = T_{max} = \dfrac{1}{2}m(v_{max})^2 = \dfrac{1}{2}(0.200)(0.77)^2 = 0.061$ J.

(d) For a bob mass of $M = 0.400$ kg, maximum angular velocity and maximum linear speed are the same as before. Total energy $= T_{mac} = \dfrac{1}{2}(0.400)(0.77)^2 = 0.12$ J.

25. $T = 2\pi\sqrt{\dfrac{I}{mgd}} = 2\pi\sqrt{\dfrac{l}{g}}$,

So, $l = \dfrac{T^2 g}{4\pi^2} = \dfrac{1.6^2 \times 9.8}{4\pi^2} = 0.64$ m (from axis of rotation).

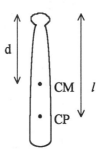

31. Letting displacement, $x(t) = A\cos(\omega t)$.

Then, $v(t) = -A\omega\sin(\omega t) = -A\omega\sqrt{1 - \cos^2(\omega t)} = -\omega\sqrt{A^2 - A^2\cos^2\omega t} = -\omega\sqrt{A^2 - x^2}$.

Thus, at $x = 1.0$ m, $v_1 = -\omega\sqrt{A^2 - 1} = 5.0$ m/s,

and, at $x = 1.5$ m, $v_2 = -\omega\sqrt{A^2 - (1.5)^2} = 3.0$ m/s.

Solving: $A = 1.72$ m, $\omega = 3.57$ rad/s, So, $T = \dfrac{2\pi}{\omega} = 1.76$ s.

37. $\Sigma F_y = ma_y$, $\Rightarrow$ $Mg - T = Ma$.

$\Sigma\tau = I\alpha$, $\Rightarrow$ $TR - kx'R = I\dfrac{a}{R}$,

So, $Mg - kx' - \left(\dfrac{I}{R^2}\right)a = Ma$, $\quad a = \dfrac{Mg - kx'}{M + I/R^2}$.

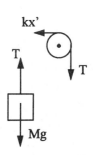

At equilibrium, $x' = l$. So relative to equilibrium $x' = x + l$

and $Mg = kl$ So, $a = \dfrac{Mg - k(x + l)}{M + I/R^2} = \dfrac{-kx}{M + I/R^2}$.

$f = \dfrac{1}{2\pi}\sqrt{\dfrac{k}{M + I/R^2}} = \dfrac{1}{2\pi}\sqrt{\dfrac{0.20}{(0.50) + (0.60)(5.0)R^2/R^2}} = 0.38$ Hz.

Chapter 15 Fluid Mechanics

1. Since mass = density × volume, $m = (2.70 \text{ gm/cm}^3)(10.0 \times 8.0 \times 30.0) = 6480 \text{ gm} = 6.48 \text{ kg}$.

7. (a) Gauge pressure $= \dfrac{\text{Force}}{\text{Area}} = \dfrac{5.0 \text{ N}}{3.0 \times 10^{-4} \text{ m}^2} = 1.67 \times 10^4 \text{ N/m}^2$.

   (b) Force $= PA = (1.67 \times 10^4)(6.0 \times 10^{-3} \times 10^{-4} \text{ m}^2) = 1.0 \times 10^{-2} \text{ N}$.

   (c) Pressure $= \rho g h = (13.6 \times 10^3 \text{ kg/m}^3)(9.8)(100 \times 10^{-3} \text{ m}) = 1.33 \times 10^4 \text{ N/m}^2$.

   Force $= (1.33 \times 10^4 \text{ N/m}^2)(3.0 \times 10^{-4} \text{ m}^2) = 4.0 \text{ N}$.

13. (a) $P = P_{atm} + \rho g h = 1.01 \times 10^5 + (1.03 \times 10^3)(9.80)(10.918) = 1.1 \times 10^8 \text{ N/m}^2$

    (b) $F = PA = (1.1 \times 10^8)(2) = 2.2 \times 10^8 \text{ N}$.   Weight of diver, $W = mg = (80)(9.8) = 784 \text{ N}$.

    Thus, the force on the window is $\dfrac{2.2 \times 10^8}{784} = 2.8 \times 10^5$ times the weight of the diver.

19. Applied force (downward) = Buoyant force (upward) = weight of water displaced.

    Thus, $F = \rho V g = (1000)\left(\dfrac{4\pi}{3}\right)(0.25)^3(9.8) = 641 \text{ N}$.

25.

At equilibrium:

$F_{B(wood)} + F_{B(lead)} = m_{wood} g + M_{lead} g$,

$\rho_{H2O} g V_{wood} + \rho_{H2O} g V_{lead} = \rho_{wood} V_{wood}\, g + \rho_{lead} V_{lead}\, g$,

$V_{lead} = V_{wood} \dfrac{\rho_{wood} - \rho_{H2O}}{\rho_{H2O} - \rho_{lead}}$

$= (8.0 \times 10^3 \text{ cm}^3)\left(\dfrac{0.60 - 1.00}{1.00 - 11.3}\right) = 310 \text{ cm}^3$.

(Wood — with forces F_{Bw} and mg; Lead — with forces F_{Bl} and Mg)

31. From equation of continuity: $v_1 A_1 = v_2 A_2 + v_3 A_3$,

    since $A_2 = A_3$; and $v_2 = v_3$.

    $v_1 A_1 = 2 v_2 A_2$,   or,

    $v_2 = v_3 = \dfrac{1}{2}\dfrac{v_1 A_1}{A_2} = \dfrac{1}{2}\dfrac{(5.0 \text{ m/s})(10.0 \text{ cm})^2(\pi/4)}{(6.0 \text{ cm})^2(\pi/4)} = 6.94 \text{ m/s}$.

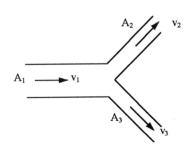

37. Equation of continuity: $v_1 A_1 = v_2 A_2$,   or, $(2.0)(\pi/4)(0.10)^2 = v_2 (\pi/4)(0.05)^2$,

    $v_2 = (2.0)\left(\dfrac{0.10}{0.05}\right)^2 = 8.0 \text{ m/s}$.   Bernoulli's equation, horizontal pipe, $P_1 + \dfrac{1}{2}\rho v_1^2 = P_2 + \dfrac{1}{2}\rho v_2^2$,

    $(0.75 \text{ atm})(1.013 \times 10^5 \text{ N/m}^2/\text{atm}) + \dfrac{1}{2}(10^3 \text{ kg/m}^3)(2.0 \text{ m/s})^2 = P_2 + \dfrac{1}{2}(10^3 \text{ kg/m}^3)(8.0 \text{ m/s})^2$,

    $P_2 = 4.6 \times 10^4 \text{ N/m}^2 = 0.45 \text{ atm}$.

43. $Q = vA$, $\Rightarrow$ $v = \dfrac{(5.0 \times 10^{-2} \text{ m}^3/\text{min})(1/60 \text{ min/s})}{(\pi/4)(0.01 \text{ m})^2} = 10.6 \text{ m/s}.$

From Bernoulli's equation: $v = \sqrt{2gh}$, (as in Problem 15-41.) So, $h = \dfrac{10.6^2}{2(9.8)} = 5.7 \text{ m}.$

49. Bernoulli's equation, $P_1 + \dfrac{1}{2}\rho v_1^2 + = P_2 + \dfrac{1}{2}\rho v_2^2$,

Since $v_2 = 0$, $\Delta P = P_1 - P_2 = -\dfrac{1}{2}\rho v_1^2 = -\dfrac{1}{2}(1.3 \text{ kg/m}^3)v_1^2$,

But $\Delta P = \rho gh = (10^3)(9.8)(10^{-2}) = 98 \text{ N/m}^2.$

Thus, $v_1^2 = \dfrac{2\Delta P}{1.3} = \dfrac{(2)(98)}{1.3} = 151$, and $v_1 = 12.3 \text{ m/s}.$

55. (a) At 20°C, $\eta_{H2O} = 1.01 \times 10^{-3} \text{ N·s/m}^2.$

From Poiseuille's Law: $(P_1 - P_2) = \dfrac{8\eta LQ}{\pi R^4} = \dfrac{8(1.01 \times 10^{-3} \text{ N·s/m}^2)(10^3 \text{ m})(0.10 \text{ m}^3/\text{s})}{\pi(0.050 \text{ m})^4} = 4.1 \times 10^4 \text{ N/m}^2.$

(b) At 0°C, $\eta_{H2O} = 1.79 \times 10^{-3} \text{ N·s/m}^2.$

$(P_1 - P_2) = \dfrac{8(1.79 \times 10^{-3} \text{ N·s/m}^2)(10^3 \text{ m})(0.10 \text{ m}^3/\text{s})}{\pi(0.050 \text{ m})^4} = 7.3 \times 10^4 \text{ N/m}^2.$

61. With zero of potential energy corresponding to columns L/2 high on each side.

$U = (\rho gAx)\dfrac{x}{2} + \left[\rho g(L-x)A\right]\left(\dfrac{L-x}{2}\right) - 2\left(\rho gA\dfrac{L}{2}\right)\dfrac{L}{4} = \rho gA\left(x^2 - Lx + \dfrac{L^2}{4}\right) = \rho gA(x - L/2)^2.$

$KE = \dfrac{1}{2}\rho ALv^2$, $E = KE + U = \dfrac{1}{2}\rho Alv^2 + \dfrac{1}{2}(2\rho gA)(x - L/2)^2.$

With $KE = \dfrac{1}{2}mv^2$ and $U = \dfrac{1}{2}kx^2$, and $T = 2\pi\sqrt{\dfrac{m}{k}}$,

$T = 2\pi\sqrt{\dfrac{\rho AL}{2\rho gA}} = 2\pi\sqrt{\dfrac{L}{2g}} = \pi\sqrt{\dfrac{2L}{g}}.$

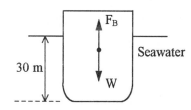

67. Let cross-section area of ship at waterline be A.

Then, since buoyant force = weight.

$F_B = W = (30A)(\rho_{seawater})$, and

$F_B = W = (dA)(\rho_{freshwater})$.

or, $dA\rho_{freshwater} = 30A\rho_{seawater}$,

$d = (30 \text{ m})\left(\dfrac{1.025}{1.000}\right) = 30.8 \text{ m}.$

Chapter 16 Waves

1. $F_1(x - 3.0t)$, $v_1 = 3.0$ m/s.

    $F_2(x + 2.0t)$, $v_2 = -2.0$ m/s.

    $F_3(y + 5.0t)$, $v_3 = -5.0$ m/s.

7. $T = \mu v^2 = (0.01 \text{ kg/m})(15 \text{ m/s})^2 = 2.25$ N.

13. $\mu = \dfrac{m}{L} = \dfrac{\rho(\pi/4)d^2L}{L} \propto d^2$.   Then, $\dfrac{v_1}{v_2} = \sqrt{\dfrac{\mu_2}{\mu_1}} = \sqrt{\left(\dfrac{d_2}{d_1}\right)^2} = \dfrac{d_2}{d_1} = 1.2$,

19. $v = \sqrt{\dfrac{B}{\rho}} = 1.45 \times 10^3$ m/s, $B = \rho(1.45 \times 10^3)^2 = (10^3 \text{ kg/m}^3)(1.45 \times 10^3)^2 = 2.10 \times 10^9 \text{ N/m}^2$.

25. (a) At $t = 0$: $y(x) = 2.0\cos\left(\dfrac{2\pi}{\lambda}x\right) = 2.0\cos\left(\dfrac{\pi x}{2}\right)$ (cgs units)

    (b) $y(x, t) = 2.0\cos\left(\dfrac{\pi x}{2} \pm \omega t\right)$.

    The expression cannot be completed without knowing the velocity and direction of travel.

31. $y(x, t) = 0.50\sin(1.2x - 100t)$ (cgs units).

Amplitude,	$A = 0.50$ cm,	Frequency,	$f = \dfrac{\omega}{2\pi} = \dfrac{100}{2\pi} = 15.9$ Hz.
Wavelength,	$\lambda = \dfrac{2\pi}{1.2} = 5.24$ cm,	Period,	$T = \dfrac{1}{f} = 0.063$ s.
Velocity,	$v = f\lambda = (15.9)(5.24) = 83.3$ cm/s.		

37. Let $y(x, t) = (0.020)\cos\left[2\pi\left(\dfrac{x}{0.40} - 25t + \dfrac{1}{6}\right)\right]$ (SI units).

    (a) $\lambda = 0.40$ m, $f = 25$ Hz, $v = f\lambda = (25)(0.40) = 10$ m/s.

    (b) $v_{max} = A\omega = A(2\pi f) = (0.020)(2\pi)(25) = 3.14$ m/s.

    $a_{max} = A\omega^2 = A(2\pi f)^2 = (0.020)(4\pi^2)(25)^2 = 493 \text{ m/s}^2$.

43. $y(x, t) = A\sin(kx - \omega t + \phi)$,   where $\lambda = \dfrac{v}{f} = \dfrac{340 \text{ m/s}}{3.0 \times 10^3 \text{ Hz}} = 0.113$ m,

    $k = \dfrac{2\pi}{\lambda} = \dfrac{2\pi}{0.113} = 55.6 \text{ m}^{-1}$, $\omega = 2\pi f = 2\pi(3.0 \times 10^3) = 1.88 \times 10^4$ rad/s.

    (a) $y(x, t) = (4.0 \times 10^{-9})\cos\left[55.6x - (1.88 \times 10^4)t\right]$ (SI units),

    $\Delta p_o = \rho v\omega A = (1.24)(340)(1.88 \times 10^4)(4.0 \times 10^{-9}) = 3.2 \times 10^{-2} \text{ N/m}^2$.

(b) $\Delta p(x, t) = (3.2 \times 10^{-2})\cos\left[55.6x - (1.88 \times 10^4)t\right]$ (SI units).

49. $m = 0.25$ kg   and $L = 3.1$ m.   So, $\mu = \dfrac{m}{L} = \dfrac{0.25}{3.1} = 8.1 \times 10^{-2}$ kg/m.

Thus, $v = \sqrt{\dfrac{T}{\mu}} = \sqrt{\dfrac{40}{8.1 \times 10^{-2}}} = 22$ m/s.   Now, $P = 80$ W $= \dfrac{1}{2}\mu v \omega^2 A^2$,

So, $\omega = \sqrt{\dfrac{(2)(80)}{(8.1 \times 10^{-2})(22)(8.2 \times 10^{-3})^2}} = 1.15 \times 10^3$ rad/s,   and, $f = \dfrac{\omega}{2\pi} = 183$ Hz.

55. From Example 16-8:

Rate at which Sun produces energy $= 4.0 \times 10^{26}$ W. Distance from Sun to earth $R = 1.5 \times 10^{11}$ m.

$I = \dfrac{P}{A} = \dfrac{4.0 \times 10^{26} \text{ W}}{4\pi(1.5 \times 10^{11} \text{ m})^2} = 1.4 \times 10^3$ W/m^2.

61. $\beta = 10 \log\left(\dfrac{I}{10^{-12}}\right) = 10 \log\left(\dfrac{2.5 \times 10^{-6}}{10^{-12}}\right) = 64$ dB.

67. $f_o = \left(1 + \dfrac{v_o}{v}\right) f_s$,   So, 450 Hz $= \left(1 + \dfrac{39}{340}\right) f_o$, and $f_o = 404$ Hz.

73. (a) $\dfrac{dT}{T} = -0.02$,   then $\dfrac{dv}{v} = \dfrac{1}{2}\left(\dfrac{dT}{T}\right) = -\dfrac{1}{2}(-0.02) = -0.01$,   or -1%.

(b) $\dfrac{dT}{T} = -0.5$,   then $\dfrac{dv}{v} = \dfrac{1}{2}\left(\dfrac{dT}{T}\right) = -\dfrac{1}{2}(-0.5) = -0.25$,   or -25%.

(c) Using Eq. 16-2: $v = \sqrt{\dfrac{T}{\mu}}$.   Let $T' = 0.5T$,

then, $v' = \sqrt{\dfrac{0.5T}{\mu}} = \sqrt{0.5}\sqrt{\dfrac{T}{\mu}} = \sqrt{0.5}\, v = 0.71v$,

or, $v' = 71\%$ of original velocity.

Part (c) is the correct result. Differentials can be used to represent small changes.

79. (a) Relate to air, whose speed is 50 km/hr $= 13.9$ m/s.

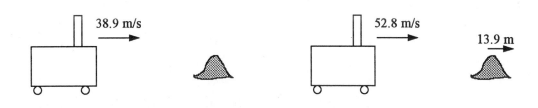

$f_o = \left(\dfrac{v + v_o}{v + v_s}\right) f_s = \left(\dfrac{340 + 52.8}{340 + 13.9}\right)(404) = 448$ Hz.

(b) Now relate to the air

$$f_o = \left(\frac{340 + 25}{340 - 13.9}\right)(404) = 452 \text{ Hz.}$$

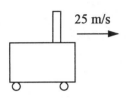

25 m/s

13.9 m/s

Chapter 17 Superposition of Waves

1.

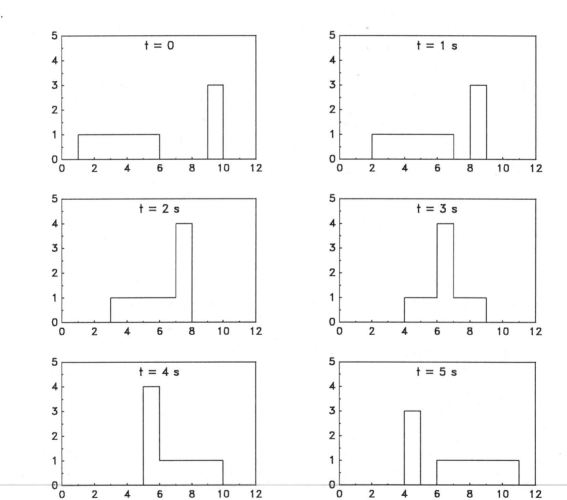

7.   From Eq. 17-2, $A = 2A_o \cos\left(\frac{\phi}{2}\right) = (2)(2.0)\cos\left(\frac{35°}{2}\right) = 3.8 \text{ cm.}$

13. $y_R(x, t) = 3.5\sin(0.50x + 400t + \phi/2)$, where $3.5 = 2A_o\cos(\phi/2)$, or, $3.5 = (2)(2.0)\cos(\phi/2)$.

(a) $\phi = 2\cos^{-1}\left[\dfrac{3.5}{(2)(2.0)}\right] = 1.0$ rad.

(b) $y_1(x, t) = 2.0\sin(0.50x + 400t)$, $y_2(x, t) = 2.0\sin(0.50x + 400t + 1.0)$ (cgs units).

19. Since $\omega_1 = 800\pi$ rad/s and $\omega_2 = 802$ rad/s. Then, $f_1 = \dfrac{\omega}{2\pi} = \dfrac{800\pi}{2\pi} = 400$ Hz,

and $f_2 = 401$ Hz. The beat frequency $f_B = f_2 - f_1 = 401 - 400 = 1$ Hz.

25. Since $y_{in}(x, t) = (2.0\times10^{-3})\cos\pi(2.0x - 50t)$,

Also $k_2 = \dfrac{\omega}{v_2} = \dfrac{50\pi \text{ rad/s}}{50 \text{ m/s}} = \pi$ m^{-1}.

Then $v_1 = \dfrac{\omega}{k_1} = \dfrac{50\pi}{2.0\pi} = 25$ m/s.

Thus, when $v_1 = 25$ m/s $< v_2 = 50$ m/s,

$y_{tr}(0, t) = \dfrac{2v_2}{v_1 + v_2}\, y_{in}(0, t)$,

and $y_{re}(0, t) = \dfrac{v_2 - v_1}{v_1 + v_2}\, y_{in}(0, t)$. $A_{tr} = \dfrac{(2)(50)}{25 + 50}\, A_{in} = \dfrac{4}{3}(2.0\times10^{-3}) = \dfrac{8.0}{3}\times10^{-3}$ m,

and $A_{re} = \dfrac{50 - 25}{25 + 50}\, A_{in} = \dfrac{1}{3}(2.0\times10^{-3}) = \dfrac{2.0}{3}\times10^{-3}$ m.

Finally: $y_{tr}(x, t) = \left(\dfrac{8.0\times10^{-3}}{3}\right)\cos\pi(1.0x - 50t)$ (SI units)

$y_{re}(x, t) = \left(\dfrac{2.0\times10^{-3}}{3}\right)\cos\pi(2.0x + 50t)$ (SI units).

31. (a) Since $f = \dfrac{v}{\lambda}$, $\dfrac{v_{tr}}{v_{in}} = \dfrac{\lambda_{tr}}{\lambda_{in}} = 1.3$,

(b) Frequency remains constant, $f_{tr} = f_{in}$.

(c) $A_{tr} = \dfrac{2v_{tr}}{v_{tr} + v_{in}}\, A_{in} = \dfrac{2A_{in}}{1 + v_{in}/v_{tr}}$, or, $\dfrac{A_{tr}}{A_{in}} = \dfrac{2}{1 + 1/1.3} = 1.13$.

37. Distance from node to node is one-half wavelength. Thus, $\lambda = 2d = 2(0.40) = 0.80$ m.

43. For a string fixed at both ends: $f = \dfrac{nv}{2L}$, $n = 1, 2, 3, \cdots$

For $n = 2$, $f = 80$ Hz $= \dfrac{2v}{(2)(2.0)}$, $v = 160$ m/s.

49. Tension $F = mg = (2.2)(9.8) = 21.6$ N. Now, $f_1 = \dfrac{v_1}{2L}$ and $f_2 = \dfrac{v_2}{2L}$.

Thus, $\dfrac{f_1}{f_2} = \dfrac{v_1}{v_2} = \sqrt{\dfrac{F_1}{F_2}}$. or, $F_2 = F_1\left(\dfrac{v_2}{v_1}\right)^2 = F_1\left(\dfrac{f_2}{f_1}\right)^2 = (21.6)\left(\dfrac{260}{220}\right)^2 = 30.2$ N.

But $\quad$ $F_2 = (2.2 + M)(9.8) = 30.2$ N, $\qquad$ So, $\qquad$ M = 0.88 kg.

55. $\quad$ $\lambda = 4L = \dfrac{v}{f}$,

$\quad$ $f = \dfrac{v}{4L} = \dfrac{340 \text{ m/s}}{(4)(0.47 \text{ m})} = 181$ Hz.

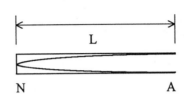

61. $\quad$ For a tube closed at one end: $\qquad$ $\lambda = \dfrac{4L}{n}$, $\qquad$ $n = 1, 3, 5, \cdots$,

$\quad$ or, $\qquad$ $f = \dfrac{v}{\lambda} = \dfrac{nv}{4L}$, $\qquad$ $n = 1, 3, 5, \cdots$

$\quad$ Then, $\quad$ $f_n = \dfrac{nv}{4L} = 300$ Hz, $\qquad$ $f_{n+2} = \dfrac{(n+2)v}{4L} = 500$ Hz.

$\quad$ Dividing: $\qquad$ $\dfrac{n+2}{n} = \dfrac{500}{300} = \dfrac{5}{3}$, $\qquad$ or, $\qquad$ $5n = 3(n+2) = 3n + 6$, $\quad$ n = 3.

$\quad$ $L = \dfrac{3v}{4f_3} = \dfrac{3(340 \text{ m/s})}{4(300 \text{ Hz})} = 0.85$ m.

67. $\quad$ For a string fixed at both ends: $\qquad$ $f = \dfrac{v}{2L}$. $\qquad$ Thus,

$\quad$ $v_1 = 2f_1 L = (2)(196)L$, $\qquad$ $v_2 = 2f_2 L = (2)(294)L$, $\qquad$ $v_3 = 2f_3 L = (2)(440)L$, $\qquad$ $v_4 = 2f_4 L = (2)(659)L$.

$\quad$ or, $\qquad$ $\dfrac{v_2}{v_1} = \dfrac{294}{196} = 1.50$, $\qquad$ $\dfrac{v_3}{v_1} = \dfrac{440}{196} = 2.24$, $\qquad$ $\dfrac{v_4}{v_1} = \dfrac{659}{196} = 3.36$.

$\quad$ Also, $\quad$ since $v = \sqrt{\dfrac{F}{\mu}}$, $\qquad$ or, $\qquad$ $\mu = \dfrac{F}{v^2}$,

$\quad$ $\dfrac{\mu_2}{\mu_1} = \left(\dfrac{v_1}{v_2}\right)^2 = \left(\dfrac{1}{1.50}\right)^2 = 0.44$. $\qquad$ $\dfrac{\mu_3}{\mu_1} = \left(\dfrac{v_1}{v_3}\right)^2 = \left(\dfrac{1}{2.24}\right)^2 = 0.20$. $\qquad$ $\dfrac{\mu_4}{\mu_1} = \left(\dfrac{v_1}{v_4}\right)^2 = \left(\dfrac{1}{3.36}\right)^2 = 0.09$.

73. $\quad$ (a) $v_{in} = \lambda_1 f_1 = (0.50 \text{ m})(200 \text{ Hz}) = 100$ m/s.

$\quad$ $v_{tr} = \sqrt{\dfrac{F}{\mu_2}} = \sqrt{\dfrac{F}{2\mu_1}} = \dfrac{1}{\sqrt{2}} v_{in}$

$\quad$ $= \dfrac{100}{\sqrt{2}}$ m/s = 71 m/s.

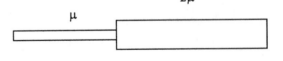

$\quad$ $f_2 = 200$ Hz, $\qquad$ So, $\qquad$ $\lambda_2 = \dfrac{v_{tr}}{f_2} = \dfrac{71}{200} = 0.35$ m.

$\quad$ (b) $k_1 = \dfrac{2\pi}{\lambda_1} = \dfrac{2\pi}{0.50 \text{ m}} = 4.0\pi$ m^{-1}, $k_2 = \dfrac{2\pi}{\lambda_2} = \dfrac{2\pi}{0.35 \text{ m}} = 5.7\pi$ m^{-1}.

$\quad$ $A_{tr} = \dfrac{2v_2}{v_1 + v_2} A_{in} = \dfrac{2(71)}{171} A_{in} = 0.083$ m. $\qquad$ $A_{re} = \dfrac{v_2 - v_1}{v_1 + v_2} A_{in} = \dfrac{71 - 100}{171} A_{in} = -0.017$ m.

$\quad$ $y_{in} = (0.10)\cos\pi(4.0x - 400t)$ (SI units), $\qquad$ $y_{tr} = (0.083)\cos\pi(5.7x - 400t)$ (SI units),

$\quad$ and $\qquad$ $y_{re} = -(0.017)\cos\pi(4.0x + 400t)$ (SI units).

79.   In air: $\lambda = 2L = \dfrac{v}{f}$,

$$L = \frac{v}{2f} = \frac{340 \text{ m/s}}{(2)(80 \text{ Hz})} = 2.13 \text{ m}.$$

In gas, $\lambda = \dfrac{2L}{3} = \dfrac{v}{f}$,

$$v = \frac{2Lf}{3} = \frac{(2)(2.13)(330 \text{ Hz})}{3} = 468 \text{ m/s}.$$

$$v = \sqrt{\frac{B}{\rho}} = 468 \text{ m/s}, \qquad B = \rho(468)^2 = (1.0)(468)^2 = 2.2 \times 10^5 \text{ N/m}^2.$$

Chapter 18 Temperature and the Properties of Gases

1.   (a) Each Celsius division equals $\dfrac{180}{100} = \dfrac{9}{5}$ Fahrenheit division. Thus, 100 Fahrenheit divisions equal $\left(\dfrac{5}{9}\right)(100) = 55.5$ Celsius divisions.

(b) Since the divisions on the Celsius and Kelvin scales are identical, 100 Fahrenheit divisions equal 55.5 Kelvin divisions.

7.   Using the results of Problem 18-6: $T_H = 4T_C + 800$, $T_H = 2.22T_F + 729$, $T_H = 4T_K - 292$,

(a) $T_H = 4(74) + 800 = 1096°H$.   (b) $T_H = 2.22(-40) + 729 = 640°H$.   (c) $T_H = 4(0) - 292 = -292°H$.

13. $(1 \text{ L·atm})(10^{-3} \text{ m}^3/\text{L})(1.013 \times 10^5 \text{ N/m}^2/\text{atm}) = 101.3 \text{ J}$.

Thus, $R = (0.08207 \text{ L·atm/mol·K})(101.3 \text{ J/L·atm}) = 8.314 \text{ J/mol·K}$.

19.   For H_2 : $P_{H2} = \dfrac{nRT}{V} = \dfrac{(0.050)(0.08207)(293)}{10} = 0.12 \text{ atm}$.

For He: $P_{He} = \dfrac{(0.030)(0.08207)(293)}{10} = 0.072 \text{ atm}$.

Net pressure, $P = P_1 + P_2 = 0.12 + 0.072 = 0.192 \text{ atm}$.

25. $nT = \dfrac{PV}{R} = \text{constant}$,   thus, $n_1 T_1 = n_2 T_2$,   or, $n_2 = \left(\dfrac{300}{200}\right)n_1 = 1.5n_1$.

Since number of moles of air in the flask increases, air is moved into the flask.

Fractional change in mass in flask: $\dfrac{m_2 - m_1}{m_1} = \dfrac{n_2 - n_1}{n_1} = \dfrac{n_2}{n_1} - 1 = 1.5 - 1 = 0.5$ or 50%.

31. $\left(\frac{1}{2}m\overline{v^2}\right) = \frac{3}{2}kT = \frac{3}{2}(1.38\times10^{-23} \text{ J/K})(300 \text{ K}) = 6.2\times10^{-21} \text{ J}$,

$\Delta U = mgh = (2 \text{ amu})(1.66\times10^{-27} \text{ kg/amu})(9.80 \text{ m/s}^2)(1.0 \text{ m}) = 3.3\times10^{-26} \text{ J}$.

Thus, gravitational potential energy is negligible when compared with average molecular kinetic energy:

$\frac{\Delta U}{K} = \frac{3.3\times10^{-26}}{6.2\times10^{-21}} = 5\times10^{-6}$.

37. Apply the Impulse = Change in momentum in a direction normal to plate:

$\overline{F}t = p_f - p_i = 2p_i = 2\Sigma(mv_x) = (2)(100)(10^{-3} \text{ kg})(100 \text{ m/s})\cos60° = 10 \text{ N·s}$.

$\overline{F} = \frac{10 \text{ N·s}}{10^{-3} \text{ s}} = 10^4 \text{ N}$, $\qquad P = \frac{\overline{F}}{A} = \frac{10^4 \text{ N}}{2.0\times10^{-4} \text{ m}^2} = 5.0\times10^7 \text{ N/m}^2$.

43. $\frac{V}{T} = \frac{nR}{P} = \text{constant}$, $\qquad \frac{V_1}{T_1} = \frac{V_2}{T_2}$, $\qquad T_2 = T_1\left(\frac{V_2}{V_1}\right) = (300 \text{ K})\left(\frac{55A}{40A}\right) = 413 \text{ K}$.

49. Outside: $\qquad 50\% = \frac{P_{partial}}{4.579}\times100\%$, $\qquad P_{partial} = 2.29 \text{ mm Hg}$.

At 20°C, $\qquad \text{R.H.} = \frac{2.29}{17.535}\times100\% = 13\%$.

Chapter 19 Thermal Properties of Matter

1. $KE_1 = \frac{3}{2}kT = \frac{3}{2}(1.38\times10^{-23} \text{ J/K})(300 \text{ K}) = 6.2\times10^{-21} \text{ J}$.

7. (a) $E = m(1.2 \text{ W/kg})(t) = (70 \text{ kg})(1.2 \text{ W/kg})(3600 \text{ s}) = 3.0\times10^5 \text{ J}$.

(b) $(150 \text{ Cal})(4186 \text{ J/Cal}) = 6.3\times10^5 \text{ J}$. The young man will have to rest for about 2 hours to burn the calorie content of the doughnut. Exercise will increase the rate of consuming energy.

13. The molar heat capacity $C = \frac{Q}{n\Delta T}$, where $n = \frac{m}{M}$.

Thus, $C = \frac{MQ}{m\Delta T} = Mc$, where M is the molar mass.

(a) For copper, $\qquad c = 390 \text{ J/kg·°C, and} \qquad M = 64 \text{ gm/mol, } C = (0.064)(390) = 25 \text{ J/mol·°C}$.

(b) For aluminum, $\qquad c = 900 \text{ J/kg·°C, and} \qquad M = 27 \text{ gm/mol, } C = (0.027)(900) = 24 \text{ J/mol·°C}$.

(c) For NaCl, $\qquad c = 880 \text{ J/kg·°C, and} \qquad M = 58 \text{ gm/mol, } C = (0.058)(880) = 51 \text{ J/mol·°C}$.

19. If all of the ice melts, heat absorbed: $\qquad Q_1 = mL_f = (50 \text{ gm})(80 \text{ cal/gm}) = 4000 \text{ cal}$.

Heat released by condensing vapor: $Q_2 = mL_v = (10 \text{ gm})(539 \text{ cal/gm}) = 5390 \text{ cal.}$

If all of the melted ice water is heated to 100°C, $Q_3 = mc\Delta T = (50)(1.0)(100) = 5000 \text{ cal.}$

Hence, enough heat is released by condensing vapor to melt all of the ice and raise the temperature of the resulting water to a final temperature T. Heat lost = Heat gained.

$Q_2 + (10 \text{ gm})(1.0 \text{ cal/gm})(100 - T) = Q_1 + (50 \text{ gm})(1.0 \text{ cal/gm})(T - 0),$

or, $5390 + 1000 - 10T = 4000 + 50 T,$ $T = 39.8°C.$

25. $$\frac{dQ}{dt} = \kappa(T_2 - T_1)\frac{A}{L},\qquad \kappa = \frac{(dQ/dt)L}{A(T_2 - T_1)} = \frac{(30 \text{ W})(0.020 \text{ m})}{(0.10 \text{ m}^2)(10°C)} = 0.60 \text{ W/m}\cdot°C.$$

31. $$L = L_o(1 + \alpha\Delta T),\qquad L_o = \frac{L}{1 + \alpha\Delta T} = \frac{2.00 \text{ m}}{1 + (20\times10^{-6}\text{ /°C})(100°C)} = 1.996 \text{ m.}$$

37. For a physical pendulum, $\tau = 2\pi\sqrt{\dfrac{I}{mgL}}$, where I is moment of inertia about pivot and L is distance from pivot to center of mass.

$$\tau + \Delta\tau = 2\pi\sqrt{\frac{I + \Delta I}{mg(L + \Delta L)}} = 2\pi\sqrt{\frac{I}{mgL}}\sqrt{\frac{1 + \Delta I/I}{1 + \Delta L/L}} \approx \tau\left(1 + \frac{\Delta I}{2I}\right)\left(1 - \frac{\Delta L}{2L}\right)$$

$$\approx \tau\left(1 + \frac{\Delta I}{2I} - \frac{\Delta L}{2L}\right) = \tau\left(1 + \frac{2\alpha\Delta T}{2} - \frac{\alpha\Delta T}{2}\right) = \tau\left(1 + \frac{\alpha}{2}\Delta T\right). \qquad \text{So,} \qquad \Delta\tau = \tau\frac{\alpha}{2}\Delta T.$$

43. Heat loss through window, $$\frac{dQ_{Win}}{dt} = \frac{\kappa A\Delta T}{L} = \frac{(0.8 \text{ W/m}\cdot C°)(2.0\times1.0 \text{ m}^2)(\Delta T)}{5.0\times10^{-3} \text{ m}} = 320(\Delta T) \text{ W.}$$

Heat loss through red brick wall, $$\frac{dQ_{brick}}{dt} = \frac{(0.6 \text{ W/m}\cdot C°)(4\times8 - 2.0)(\Delta T)}{0.10 \text{ m}} = 180(\Delta T) \text{ W.}$$

Ratio, $$\frac{dQ_{win}/dt}{dQ_{brick}/dt} = \frac{320(\Delta T)}{180(\Delta T)} = 1.8.$$

49. For glass bottle, $\Delta V_g = V_o(\beta_g)\Delta T = (1 \text{ L})(1.2\times10^{-5}\text{ /°C})(30°C - 20°C) = 1.2\times10^{-4} \text{ L.}$

For water, $\Delta V_w = V_o(\beta_w)\Delta T = (1 \text{ L})(25\times10^{-5}\text{ /°C})(30°C - 20°C) = 2.5\times10^{-3} \text{ L.}$

Amount of overflow, $\Delta V = (2.5 - 0.1)\times10^{-3} = 2.4 \text{ mL.}$

55. Heat required to raise temperature of two moles from 30 to 80 K is approximately 1.7×10^3 J.

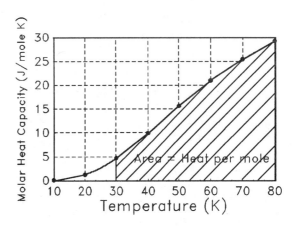

Chapter 20 First Law of Thermodynamics

1. $W = p(V_2 - V_1) = (2.0 \text{ atm})(5.0 - 3.0 \text{ L}) = 4 \text{ L·atm} = 404 \text{ J}.$

7. Path AB: constant pressure, $W = \text{Area} = p(V_2 - V_1) = (1 \text{ atm})(3 - 1 \text{ L}) = 2 \text{ L·atm} = 202 \text{ J}.$

    Path ADB: $W = \text{Area} = (1 \text{ atm})(2 \text{ L}) + \frac{1}{2}(2 \text{ atm})(2 \text{ L}) = 4 \text{ L·atm} = 404 \text{ J}.$

    Path ACB: $W = (1 \text{ atm})(2 \text{ L}) + \frac{1}{2}(2 \text{ atm})(2 \text{ L}) = 4 \text{ L·atm} = 404 \text{ J}.$

    Path ADCB: $A = (3 \text{ atm})(2 \text{ L}) = 6 \text{ L·atm} = 606 \text{ J}.$

13. From Problem 20-6: $\Delta U = 0$, (isothermal),   and $W = pV\ln(4)$.

    Thus, $Q = \Delta U + W = pV\ln(4)$.

19. From A to B: $\Delta U_{AB} = Q - W = 250 - 500 = -250 \text{ J}.$

    From A to C: $\Delta U_{AC} = Q - W = 300 - 700 = -400 \text{ J}.$

    (a) From B to C: $\Delta U_{BC} = \Delta U_{AC} - \Delta U_{AB} = -400 - (-250) = -150 \text{ J}.$

    But, $\Delta U_{BC} = Q - W = Q - (0) = -150 \text{ J},$ or, $Q = -150 \text{ J}.$

    (b) From C to D to A: $\Delta U_{CA} = -\Delta U_{AC} = 400 \text{ J}.$

    But, $\Delta U_{CA} = Q - W = Q - (-800) = 400 \text{ J},$ or, $Q = 400 - 800 = -400 \text{ J}.$

25. $\Delta U = mgy = m_{ice}L_f,$ $y = \dfrac{(2.0\times10^{-4} \text{ kg})(80 \text{ cal/gm})(4.186 \text{ J/cal})(10^3 \text{ gm/kg})}{(1.0 \text{ kg})(9.80 \text{ m/s}^2)} = 6.8 \text{ m}.$

31. $nRT = pV = (0.50 \text{ atm})(50 \text{ L})(1.013\times10^5 \text{ N/m}^2/\text{atm})(10^{-3} \text{ m}^3/\text{L}) = 2530 \text{ J}.$

    For polyatomic gas $U = 3nRT = 3(pV) = 7600 \text{ J}.$

37. For ideal gas, adiabatic process with volume and temperature:

    $T_1V_1^{\gamma-1} = T_2V_2^{\gamma-1}$, where $\gamma = 1.67$.

    $T_2 = T_1 \left(\dfrac{V_1}{V_2}\right)^{\gamma-1} = (300 \text{ K})\left(\dfrac{1}{2}\right)^{0.67} = 189 \text{ K}.$

43. (a) $T = \dfrac{pV}{nR} = \dfrac{(2.0\times10^5 \text{ N/m}^2)(1.0\times10^{-2} \text{ m}^2)}{(1.0 \text{ mol})(8.314 \text{ J/mol·K})} = 241 \text{ K}.$

    (b) $T_2 = T_1 \left(\dfrac{V_1}{V_2}\right)^{\gamma-1} = (241 \text{ K})\left(\dfrac{1.0\times10^{-2}}{5.0\times10^{-3}}\right)^{0.67} = 383 \text{ K}.$

    (c) From Problem 20-40, $W = \dfrac{nR}{\gamma-1}\left(T_1 - T_2\right) = \dfrac{(1.0 \text{ mol})(8.314 \text{ J/mol·K})}{0.67}(241 - 383 \text{ K}) = -1760 \text{ J}.$

    So 1760 J done on the gas.   (d) $\Delta U = Q - W = 0 - (-1760) = 1760 \text{ J}.$

Chapter 21 The Second Law of Thermodynamics

1. $\varepsilon = \dfrac{W}{Q_h}$, $\Rightarrow$ $Q_h = \dfrac{W}{\varepsilon} = \dfrac{200 \text{ J}}{0.40} = \boxed{500 \text{ J}}$, $Q_c = Q_h - W = 500 \text{ J} - 200 \text{ J} = \boxed{300 \text{ J}}$.

7. (a) $\kappa = \dfrac{Q_c}{Q_h - Q_c}$, $\Rightarrow$ $Q_c = \dfrac{\kappa Q_h}{1 + \kappa} = \dfrac{(6.0)(80 \text{ J})}{1 + 6.0} = \boxed{69 \text{ J}}$,

    (b) $W = Q_h - Q_c = 80 \text{ J} - 68.6 \text{ J} = \boxed{11 \text{ J}}$.

13. $\dfrac{Q_c}{Q_h} = \dfrac{T_c}{T_h} = \dfrac{300 \text{ K}}{600 \text{ K}} = 0.50$, $Q_c = (0.50)(100 \text{ J}) = 50 \text{ J}$, $W = Q_h - Q_c = 100 \text{ J} - 50 \text{ J} = \boxed{50 \text{ J}}$.

19. $\varepsilon = 1 - \dfrac{T_c}{T_h} = 1 - \dfrac{293 \text{ K}}{473 \text{ K}} = \boxed{0.38}$. So 45% is impossible, and 35% is feasible.

25. For isothermal process, $Q = W = 20 \text{ J}$, So, $\Delta S = \dfrac{Q}{T} = \dfrac{20 \text{ J}}{300 \text{ K}} = \boxed{0.067 \text{ J/K}}$.

31. (a) $Q = W = nRT\ln(\dfrac{V_2}{V_1})$, $\Delta S = \dfrac{Q}{T} = nR\ln(\dfrac{V_2}{v_1}) = (1.0 \text{ mol})(8.31 \text{ J/mol·K})\ln(2.0) = \boxed{5.8 \text{ J/K}}$,

    (b) $Q = W = 1500 \text{ J} = nRT\ln(\dfrac{V_2}{V_1})$, $T = \dfrac{1500 \text{ J}}{(1.0 \text{ mol})(8.31 \text{ J/mol·K})\ln(2.0)} = \boxed{260 \text{ K}}$.

37. From 0°C ice to 0°C water, $Q_1 = mL_f = (0.050 \text{ kg})(3.35 \times 10^5 \text{ J/kg}) = 1.68 \times 10^4 \text{ J}$,

    $\Delta S_1 = \dfrac{Q}{T} = \dfrac{1.68 \times 10^4 \text{ J}}{273 \text{ K}} = 61.5 \text{ J/K}$, From 0°C water to 100°C water, $dQ_2 = mcdT$,

    $\Delta S_2 = \displaystyle\int_{T_1}^{T_2} \dfrac{dQ}{T} = \int_{T_1}^{T_2} \dfrac{mcdT}{T} = mc\ln(\dfrac{T_2}{T_1}) = (0.050 \text{ kg})(4186 \text{ J/kg·C°})\ln(\dfrac{373}{273}) = 65.3 \text{ J/K}$,

    From 100°C water to 100°C steam, $Q_3 = mL_v = (0.050 \text{ kg})(2.26 \times 10^6 \text{ J/kg}) = 1.13 \times 10^5 \text{ J}$,

    $\Delta S_3 = \dfrac{1.13 \times 10^5 \text{ J}}{373 \text{ K}} = 303 \text{ J/K}$, $\Delta S = \Delta S_1 + \Delta S_2 + \Delta S_3 = 61.5 \text{ J/K} + 65.3 \text{ J/K} + 303 \text{ J/K} = \boxed{430 \text{ J/K}}$.

43. (a) $A \rightarrow B$, $\Delta S = 2.0 \text{ J/K} = \dfrac{Q_h}{T_h}$, $Q_h = (2.0 \text{ J/K})(600 \text{ K}) = \boxed{1200 \text{ J}}$,

    (b) $C \rightarrow D$, $\Delta S = -2.0 \text{ J/K} = \dfrac{-Q_c}{T_c}$, $Q_c = (2.0 \text{ J/K})(300 \text{ K}) = \boxed{600 \text{ J}}$,

    (c) $W = Q_h - Q_c = 1200 \text{ J} - 600 \text{ J} = \boxed{600 \text{ J}}$,

    (d) $\varepsilon = \dfrac{W}{Q_h} = \dfrac{600 \text{ J}}{1200 \text{ J}} = \boxed{0.50}$.

49. (a) The first law, $dU = đQ - đW$,   The second law, $dS = \dfrac{đQ}{T}$, $\Rightarrow$ $đQ = TdS$,

So, $\boxed{dU = TdS - đW}$,

(b) $dS = \dfrac{dU}{T} + \dfrac{đW}{T}$,   Also $pV = nRT$,

$$\Delta S = \int \dfrac{dU}{T} + \int \dfrac{đW}{T} = \int_{T_1}^{T_2} \dfrac{nC_v dT}{T} + \int_{V_1}^{V_2} \dfrac{pdV}{T} = \int_{T_1}^{T_2} \dfrac{nC_v dT}{T} + \int_{V_1}^{V_2} \dfrac{nRdV}{V} = \boxed{nC_v\ln(\dfrac{T_2}{T_1}) + nR\ln(\dfrac{V_2}{V_1})}.$$

55. From Prob. 51, $\Delta S = -nR\ln(\dfrac{P_2}{P_1}) = -(\dfrac{500}{4}\text{ mol})(8.31\text{ J/mol·K})\ln(\dfrac{1}{120}) = \boxed{5.0\times10^3\text{ J/K}}$.

Chapter 22 Electric Charge and Coulomb's Law

1. In 1.0 mole of H_2 gas, there are 6.02×10^{23} molecules or $2(6.02\times10^{23}) = 1.20\times10^{24}$ electrons.
So the net charge $Q = (1.60\times10^{-19}\text{ C})(1.20\times10^{24}) = \boxed{1.9\times10^5\text{ C}}$.

7. $8\text{ oz} = 8(28.4\text{ g}) = 227\text{ g}$, there are $\dfrac{227\text{ g}}{18\text{ g/mol}} = 12.6\text{ mol}$ or $12.6(6.02\times10^{23}) = 7.59\times10^{24}$ molecules.

In each molecule, there are 10 electrons and 10 protons, so the charge of all the electrons is

$q_1 = 10(7.59\times10^{24})(-1.60\times10^{-19}\text{ C}) = -1.21\times10^7\text{ C}$ and the charge of all the nuclei is $q_2 = 1.21\times10^7\text{ C}$.

$F = k\dfrac{q_1 q_2}{r^2} = (9.0\times10^9\text{ N·m}^2/\text{C}^2)\times\dfrac{(1.21\times10^7\text{ C})^2}{(6.0/3.28\text{ m})^2} = 3.94\times10^{23}\text{ N} = \boxed{8.9\times10^{22}\text{ lb}}$.

The $-$ sign indicates an attractive force.

13. In y direction, $T\cos5.0° = mg$,

In x direction, $T\sin5.0° = F_e = k\dfrac{q_1 q_2}{r^2} = k\dfrac{q^2}{r^2}$,

$F_e = mg\tan5.0° = (0.0050\text{ kg})(9.80\text{ m/s}^2)\tan5.0° = 4.29\times10^{-3}\text{ N}$

$r = 2L\sin5.0° = 2(0.50\text{ m})\sin5.0° = 0.0872\text{ m}$,

Therefore, $q = r\sqrt{\dfrac{F_e}{k}}$

$= (0.0872\text{ m})\sqrt{\dfrac{4.29\times10^{-3}\text{ N}}{9.0\times10^9\text{ N·m}^2/\text{C}^2}} = \boxed{6.0\times10^{-8}\text{ C}}$.

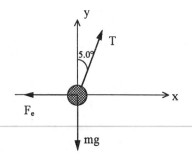

No, the sign of q cannot be determined.

19. $\mathbf{r} = \mathbf{r_2} - \mathbf{r_1} = (4.0\,\mathbf{i} + 7.0\,\mathbf{j} - 14.0\,\mathbf{k})\text{ m}$,   So, $r = \sqrt{(4.0\text{ m})^2 + (7.0\text{ m})^2 + (-14.0\text{ m})^2} = 16.2\text{ m}$,

25. $\mathbf{F} = \dfrac{kQq}{x^2}\,\mathbf{i}$, $\qquad dW = \mathbf{F}\cdot d\mathbf{x} = \dfrac{kQq\,dx}{x^2}$, $\qquad$ So, $\qquad W = \displaystyle\int_{x_1}^{x_2} \dfrac{kQq\,dx}{x^2} = -\dfrac{kQq}{x}\bigg|_{x_1}^{x_2} = \boxed{kQq\left(\dfrac{1}{x_1} - \dfrac{1}{x_2}\right)}$.

Chapter 23 The Electric Field

1. (a) $E = \dfrac{F_e}{q_t} = \dfrac{4.0\times10^{-6}\text{ N}}{2.0\times10^{-8}\text{ C}} = \boxed{200\text{ N/C upward}}$.

   (b) $F_e = qE = (-1.0\times10^{-8}\text{ C})(200\text{ N/C}) = \boxed{-2.0\times10^{-6}\text{ N}}$. The $-$ sign means the force is downward.

7. $E = k\dfrac{q}{r^2}$, $\qquad \Rightarrow \qquad q = \dfrac{Er^2}{k} = \dfrac{(100\text{ N/C})(0.50\text{ m})^2}{9.0\times10^9\text{ N·m}^2/\text{C}^2} = \boxed{2.8\times10^{-9}\text{ C}}$.

13. (a) $E_1 = k\dfrac{q_1}{r_1^{\,2}} = (9.0\times10^9\text{ N·m}^2/\text{C}^2)\times\dfrac{2.0\times10^{-7}\text{ C}}{(0.200\text{ m})^2} = 4.5\times10^4\text{ N/C}$,

   $E_2 = (9.0\times10^9\text{ N·m}^2/\text{C}^2)\times\dfrac{6.0\times10^{-8}\text{ C}}{(0.050\text{ m})^2} = 2.16\times10^5\text{ N/C}$,

   The net field $\quad E = E_1 + E_2 = \boxed{2.6\times10^5\text{ N/C}}$.

   (b) $F = eE = (1.60\times10^{-19}\text{ C})(2.6\times10^5\text{ N/C})$

   $= \boxed{4.2\times10^{-14}\text{ N toward the positive charge}}$.

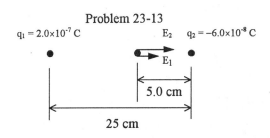

Problem 23-13

$q_1 = 2.0\times10^{-7}\text{ C}$ $\qquad E_2 \quad q_2 = -6.0\times10^{-8}\text{ C}$

5.0 cm

25 cm

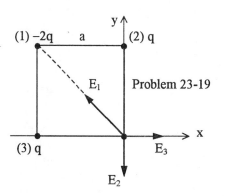

Problem 23-19

19. (a) $E_2 = E_3 = k\dfrac{q}{a^2}$, $\qquad E_1 = k\dfrac{2q}{(\sqrt{2}a)^2} = \dfrac{kq}{a^2}$,

   $\mathbf{E_1} = \dfrac{kq}{a^2}\times(-\cos45°\,\mathbf{i} + \sin45°\,\mathbf{j})$, $\qquad \mathbf{E_2} = -\dfrac{kq}{a^2}\,\mathbf{j}$, $\qquad \mathbf{E_3} = \dfrac{kq}{a^2}\,\mathbf{i}$,

   The net field $\mathbf{E} = \mathbf{E_1} + \mathbf{E_2} + \mathbf{E_3} = \boxed{(1 - \dfrac{1}{\sqrt{2}})\dfrac{kq}{a^2}(\mathbf{i} - \mathbf{j})}$,

   (b) $F = qE = \boxed{(1 - \dfrac{1}{\sqrt{2}})\dfrac{kq^2}{a^2}(\mathbf{i} - \mathbf{j})}$.

25. All charges reside on inside surface due to attraction. $\sigma = \dfrac{q}{A} = \dfrac{\pm 8.0 \times 10^{-6}\ \text{C}}{(2.0\ \text{m})^2} = \boxed{\pm 2.0 \times 10^{-6}\ \text{C/m}^2}$,

Since $2.0\ \text{m} \gg 2.0\ \text{cm}$, the plates can be considered as very large.

$E = \dfrac{\sigma}{\varepsilon_0} = \dfrac{2.0 \times 10^{-6}\ \text{C/m}^2}{8.85 \times 10^{-12}\ \text{F/m}} = \boxed{2.3 \times 10^5\ \text{N/C}}$.

31. Due to symmetry, the electric field at P is in the −y direction. The x components of the electric fields from all infinitesimal segments cancel out. The linear charge

density $\lambda = \dfrac{q}{L} = \dfrac{Q}{2R\theta}$.

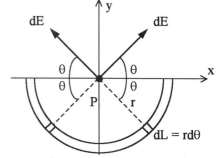

$E_y = \displaystyle\int_{\pi/2-\theta}^{\pi/2+\theta} dE_y = \dfrac{kQ}{2R^2\theta} \int_{\pi/2-\theta}^{\pi/2+\theta} \sin\alpha\, d\alpha = -\dfrac{kQ\cos\alpha}{2R^2\theta}\Big|_{\pi/2-\theta}^{\pi/2+\theta}$

$= \dfrac{kQ}{2R^2\theta} [\cos(\pi/2-\theta) - \cos(\pi/2+\theta)] = \boxed{\dfrac{kQ\sin\theta}{R^2\theta}}$.

37. (a) $F_g = mg = F_e = qE$, $\Rightarrow$ $q = \dfrac{mg}{E} = \dfrac{\rho V g}{E} = \dfrac{\rho(4\pi/3)r^3 g}{E} = \boxed{\dfrac{4\pi r^3 \rho g}{3E}}$,

(b) $q_1 = \dfrac{4\pi(1.52 \times 10^{-6}\ \text{m})^3(851\ \text{kg/m}^3)(9.80\ \text{m/s}^2)}{3(2.55 \times 10^5\ \text{N/C})} = \boxed{4.8 \times 10^{-19}\ \text{C}}$,

$q_2 = \boxed{6.4 \times 10^{-19}\ \text{C}}$, and $q_3 = \boxed{3.2 \times 10^{-19}\ \text{C}}$,

(c) The numbers of electrons are $\boxed{3}$ for q_1 , $\boxed{4}$ for q_2 , and $\boxed{2}$ for q_3 .

43. (a) $\theta = \tan^{-1}\dfrac{1}{4.73} = 11.9°$, $r = \sqrt{4.73^2 + 1^2} = 4.83\ \text{m}$.

From $-3Q$, $E_1 = k\dfrac{q}{r^2} = (9.0 \times 10^9\ \text{N·m}^2/\text{C}^2)\dfrac{3(8.0 \times 10^{-6}\ \text{C})}{(3.0\ \text{m})^2} = 2.4 \times 10^4\ \text{N/C}$,

From $2Q$, $E_2 = (9.0 \times 10^9\ \text{N·m}^2/\text{C}^2)\dfrac{2(8.0 \times 10^{-6}\ \text{C})}{[(1.0\ \text{m})/\sin 30°]^2} = 3.6 \times 10^4\ \text{N/C}$,

From Q, $E_3 = (9.0 \times 10^9\ \text{N·m}^2/\text{C}^2)\dfrac{8.0 \times 10^{-6}\ \text{C}}{(4.83\ \text{m})^2} = 3.09 \times 10^3\ \text{N/C}$,

$\mathbf{E_1} = -2.4 \times 10^4 \mathbf{i}\ \text{N/C}$,

$\mathbf{E_2} = (3.6 \times 10^4\ \text{N/C})(\cos 30°\mathbf{i} + \sin 30°\mathbf{j})$,

$\mathbf{E_3} = (3.09 \times 10^3\ \text{N/C})(\cos 11.9°\mathbf{i} + \sin 11.9°\mathbf{j})$,

The net field

$\mathbf{E} = \mathbf{E_1} + \mathbf{E_2} + \mathbf{E_3} = \boxed{(1.02\ \mathbf{i} + 1.86\ \mathbf{j}) \times 10^4\ \text{N/C}}$,

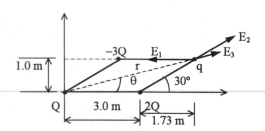

(b) $F = qE = (5.0 \times 10^6\ \text{C})[(1.02\ \mathbf{i} + 1.86\ \mathbf{j}) \times 10^4\ \text{N/C}] = \boxed{(0.051\ \mathbf{i} + 0.093\ \mathbf{j})\ \text{N}}$.

49. $$\mathbf{a} = \frac{q\mathbf{E}}{m} = \frac{(-1.60\times10^{-19}\ \text{C})[(2.0\ \mathbf{i} - 4.0\ \mathbf{j})\times10^3\ \text{N/C}]}{9.11\times10^{-31}\ \text{kg}} = (-3.51\ \mathbf{i} + 7.03\ \mathbf{j})\times10^{14}\ \text{m/s}^2 ,$$

$$\mathbf{v} = \mathbf{v_0} + \mathbf{a}t = (1.0\ \mathbf{i} - 3.0\ \mathbf{j} + 2.0\ \mathbf{k})\times10^6\ \text{m/s} + [(-3.51\ \mathbf{i} + 7.03\ \mathbf{j})\times10^{14}\ \text{m/s}^2](2.0\times10^{-8}\ \text{s})$$

$$= \boxed{(-6.0\ \mathbf{i} + 11\ \mathbf{j} + 2.0\ \mathbf{k})\times10^6\ \text{m/s}} ,$$

$$\mathbf{x} = \mathbf{v_0}t + \frac{1}{2}\mathbf{a}t^2$$

$$= [(1.0\ \mathbf{i} - 3.0\ \mathbf{j} + 2.0\ \mathbf{k})\times10^6\ \text{m/s}](2.0\times10^{-8}\ \text{s}) + \frac{1}{2}[(-3.51\ \mathbf{i} + 7.03\ \mathbf{j})\times10^{14}\ \text{m/s}^2](2.0\times10^{-8}\ \text{s})^2$$

$$= \boxed{(-5.0\ \mathbf{i} + 8.0\ \mathbf{j} + 4.0\ \mathbf{k})\ \text{cm}} .$$

55. From Problem 28, the electric field at a point x from the y axis is $E = \frac{2k\lambda_y}{x}$,

From an infinitesimal segment dx on x axis with charge $\lambda_x dx$,

the electric force is $dF = dq\cdot E = \lambda_x dx\cdot E = \frac{2k\lambda_x\lambda_y dx}{x}$,

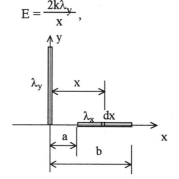

$$F = \int_a^b dF = 2k\lambda_x\lambda_y \int_a^b \frac{dx}{x} = 2k\lambda_x\lambda_y \ln x \Big|_a^b = \boxed{2k\lambda_x\lambda_y\ln(\frac{b}{a})} .$$

Chapter 24 Gauss' Law

1. $$\Phi_e = \mathbf{E}\cdot\mathbf{n}A = EA\cos\theta = (1.1\times10^4\ \text{N/C})(2.0\ \text{m})^2\cos0° = \boxed{4.4\times10^4\ \text{N}\cdot\text{m}^2/\text{C}} ,$$

7. $$\Phi_e = \oint \mathbf{E}\cdot\mathbf{n}\ da = \frac{Q_{in}}{\varepsilon_0} ,$$

For S_1 , $\Phi_e = \frac{(q - 2q - 2q + 3q)}{\varepsilon_0} = \boxed{0}$, For S_2 , $\Phi_e = \boxed{\frac{-2q}{\varepsilon_0}}$,

For S_3 , $\Phi_e = \boxed{\frac{q}{\varepsilon_0}}$, For S_4 , $\Phi_e = \frac{-2q - 2q}{\varepsilon_0} = \boxed{\frac{-4q}{\varepsilon_0}}$,

For S_5 , $\Phi_e = \boxed{\frac{-2q}{\varepsilon_0}}$, For S_6 , $\Phi_e = \boxed{\frac{3q}{\varepsilon_0}}$.

13. Still $\Phi_e = \boxed{0}$.

19. $\rho = \dfrac{q}{V} = \dfrac{-30 \times 10^{-6} \text{ C}}{4\pi(0.100 \text{ m})^3/3} = -7.16 \times 10^{-3} \text{ C/m}^3$,

(a) $Q_{in} = \rho V = (-7.16 \times 10^{-3} \text{ C/m}^3)(\dfrac{4\pi}{3})(0.020 \text{ m})^3 = -2.40 \times 10^{-7} \text{ C}$,

$\Phi_e = \displaystyle\oint \mathbf{E} \cdot \mathbf{n} \, da = EA = E(4\pi r^2) = \dfrac{Q_{in}}{\varepsilon_0}$,   So, $E = \dfrac{Q_{in}}{4\pi\varepsilon_0 r^2}$,

$E = (9.0 \times 10^9 \text{ N} \cdot \text{m}^2/\text{C}^2) \times \dfrac{-2.40 \times 10^{-7}}{(0.020 \text{ m})^2} = \boxed{-5.4 \times 10^6 \text{ N/C}}$,

(b) $Q_{in} = (-7.16 \times 10^{-3} \text{ C/m}^3)(\dfrac{4\pi}{3})(0.050 \text{ m})^3 = -3.75 \times 10^{-6} \text{ C}$,

$E = (9.0 \times 10^9 \text{ N} \cdot \text{m}^2/\text{C}^2) \times \dfrac{-3.75 \times 10^{-6} \text{ C}}{(0.050 \text{ m})^2} = \boxed{-1.4 \times 10^7 \text{ N/C}}$,

(c) $Q_{in} = -30 \times 10^{-6} \text{ C}$, $E = (9.0 \times 10^9 \text{ N} \cdot \text{m}^2/\text{C}^2) \times \dfrac{-30 \times 10^{-6} \text{ C}}{(0.200 \text{ m})^2} = \boxed{-6.8 \times 10^6 \text{ N/C}}$.

All $-$ signs mean the electric fields are toward the center of the sphere.

25. $\Phi_e = \displaystyle\oint \mathbf{E} \cdot \mathbf{n} \, da = EA = E(2\pi r L) = \dfrac{Q_{in}}{\varepsilon_0}$,   So, $E = \dfrac{Q_{in}}{2\pi\varepsilon_0 r L}$,

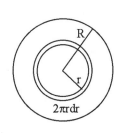

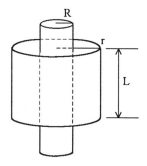

For $r \leq R$, $dQ_{in} = \rho dV = \rho d(\pi r^2 L) = 2\pi L \rho r \, dr = 2\pi L \alpha r^2 \, dr$,

$Q_{in} = \displaystyle\int_0^r dQ_{in} = 2\pi L \alpha \int_0^r r^2 \, dr = \dfrac{2\pi L \alpha r^3}{3}$,   So, $E = \boxed{\dfrac{\alpha r^2}{3\varepsilon_0} \qquad (r \leq R)}$,

For $r \geq R$, $Q_{in} = \displaystyle\int_0^R dQ_{in} = \dfrac{2\pi L \alpha R^3}{3}$,   So, $E = \boxed{\dfrac{\alpha R^3}{3\varepsilon_0 r} \qquad (r \geq R)}$.

31. First calculate the flux from charge Q. Construct a Gauss surface of a cube with side 2a centered at charge Q. $\oint \mathbf{E \cdot n} da = 6 \int \mathbf{E \cdot n}\, da = \dfrac{Q_{in}}{\varepsilon_0} = \dfrac{Q}{\varepsilon_0}$, So, $\int \mathbf{E \cdot n}\, da = \dfrac{Q}{6\varepsilon_0}$,

Due to symmetry, the $-Q$ charge will have equal contribution. Therefore the net flux $\Phi_e = \boxed{\dfrac{Q}{3\varepsilon_0}}$.

37. For $|z| \geq a/2$, $\qquad Q_{in} = \rho V = \rho Aa$,

$\Phi_e = \oint \mathbf{E \cdot n} da = E(2A) = \dfrac{Q_{in}}{\varepsilon_0} = \dfrac{\rho Aa}{\varepsilon_0}$,

So, $\qquad E = \boxed{\dfrac{\rho a}{2\varepsilon_0}} \qquad (|z| \geq a/2)$,

For $|z| < a/2$, $\qquad Q_{in} = \rho A(2z)$,

$E = \boxed{\dfrac{\rho z}{\varepsilon_0}} \qquad (|z| \leq a/2)$.

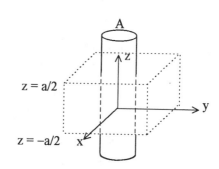

Chapter 25 Electric Potential

1. $V = k\dfrac{q}{r}$, $\qquad$ For proton, $\qquad V_p = (9.0\times 10^9 \text{ N·m}^2/\text{C}^2)\times\dfrac{1.60\times 10^{-19} \text{ C}}{0.50\times 10^{-9} \text{ m}} = \boxed{2.9 \text{ V}}$,

$\qquad\qquad\qquad\qquad$ For electron, $\qquad V_p = (9.0\times 10^9 \text{ N·m}^2/\text{C}^2)\times\dfrac{-1.60\times 10^{-19} \text{ C}}{0.50\times 10^{-9} \text{ m}} = \boxed{-2.9 \text{ V}}$.

7. $V = k\left(\dfrac{q_1}{r_1} + \dfrac{q_2}{r_2}\right)$

$= (9.0\times 10^9 \text{ N·m}^2/\text{C}^2)\left(\dfrac{8.0\times 10^{-6} \text{ C}}{0.10 \text{ m}} + \dfrac{8.0\times 10^{-6} \text{ C}}{0.10 \text{ m}}\right)$

$= 1.44\times 10^6 \text{ V}$,

$U = qV = (-4.0\times 10^{-6} \text{ C})(1.44\times 10^6 \text{ V}) = \boxed{-5.8 \text{ J}}$.

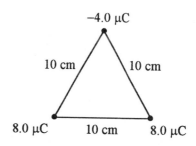

13. $V = k\left(\dfrac{q}{a/\sqrt{2}} + \dfrac{q}{a/\sqrt{2}} + \dfrac{2q}{a/\sqrt{2}}\right) = \boxed{\dfrac{4\sqrt{2}kq}{a}}$.

19. (a) $\Delta V = kq(\frac{1}{r_2} - \frac{1}{r_1}) = (9.0\times10^9 \text{ N·m}^2/\text{C}^2)(2.0\times10^{-6} \text{ C})(\frac{1}{4.0 \text{ m}} - \frac{1}{2.0 \text{ m}}) = \boxed{-4.5 \text{ kV}}$.

(b) $W_E = -q'\Delta V = -(-4.0\times10^{-6} \text{ C})(-4500 \text{ V}) = \boxed{-1.8\times10^{-2} \text{ J}}$. The – sign means we do the work.

25. $\Delta U = q\Delta V = (5.0\times10^{-6} \text{ C})(50 \text{ V}) = \boxed{2.5\times10^{-4} \text{ J}} = \boxed{1.6\times10^{15} \text{ eV}}$.

31. (a) Energy $= \Delta U = q\Delta V = (25 \text{ C})(4.0\times10^7 \text{ V}) = \boxed{1.0\times10^9 \text{ J}}$,   (b) Energy $= cm\Delta T + mL_v$,

1.0×10^9 J $= m[(4186 \text{ J/kg·C}°)(100°\text{C} - 20°\text{C}) + 2.26\times10^6 \text{ J/kg}]$,   Solve for $m = \boxed{390 \text{ kg}}$.

37. $p = 2aq = (5.0\times10^{-11} \text{ m})(1.60\times10^{-19} \text{ C}) = \boxed{8.0\times10^{-30} \text{ C·m}}$.

43. $\Delta V = E\Delta l$, $\Rightarrow$ $\Delta l = \frac{\Delta V}{E} = \frac{100 \text{ V}}{5.00\times10^3 \text{ V/m}} = \boxed{0.020 \text{ m}}$.

49. (a) $q = (1.60\times10^{-19} \text{ C})(1.0\times10^{12}) = 1.6\times10^{-7}$ C , $\sigma = \frac{q}{A} = \frac{1.6\times10^{-7} \text{ C}}{400\times10^{-4} \text{ m}^2} = \boxed{4.0\times10^{-6} \text{ C/m}^2}$,

(b) $E = \frac{\sigma}{\varepsilon_0} = \frac{4.0\times10^{-6} \text{ C/m}^2}{8.85\times10^{-12} \text{ F/m}} = \boxed{4.5\times10^5 \text{ V/m}}$,

(c) $\Delta V = E\Delta l = (4.52\times10^5 \text{ V/m})(2.0\times10^{-2} \text{ m}) = \boxed{9.0 \text{ kV}}$.

55. (a) $\sigma = \frac{q}{A} = \frac{5.0\times10^{-12} \text{ C}}{4\pi(0.035 \text{ m})^2} = \boxed{3.2\times10^{-10} \text{ C/m}^2}$,

(b) $\sigma = \varepsilon_o E = (8.85\times10^{-12} \text{ C}^2/\text{N·m}^2)(8.0 \text{ N/C}) = \boxed{7.08\times10^{-11} \text{ C/m}^2}$.

(c) $Q_{inner} = 5.0\times10^{-12}$ C,

$Q_{outer} = \sigma A = (7.08\times10^{-11} \text{ C/m}^2)(4\pi)(0.040 \text{ m})^2$

$= 1.42\times10^{-12}$ C. $Q = Q_{inner} + Q_{outer} = \boxed{6.4\times10^{-12} \text{ C}}$.

61.

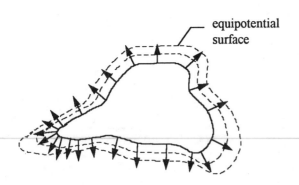

equipotential
surface

67. (a) $V = \int \frac{kdq}{r}$. From an infinitesimal ring of $dq = \sigma(2\pi r dr)$,

$$V = k\int_0^R \frac{2\pi\sigma r dr}{\sqrt{r^2 + z^2}} = 2\pi k\sigma \sqrt{r^2 + z^2} \Big|_0^R = \boxed{2\pi k\sigma (\sqrt{R^2 + z^2} - z)},$$

(b) $E_z = -\frac{dV}{dz} = \boxed{2\pi k\sigma(1 - \frac{z}{\sqrt{R^2 + z^2}})}$,

(c) The same result.

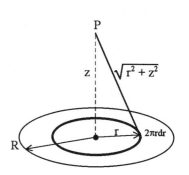

Chapter 26 Electric Current and Electromotive Force

1. $I = \frac{\Delta q}{\Delta t}$, $\Rightarrow$ $\Delta q = I\Delta t = (2.0 \text{ A})(1.0 \text{ s}) = \boxed{2.0 \text{ C}}$.

The number of electrons is $\frac{2.0 \text{ C}}{1.60\times10^{-19} \text{ C}} = \boxed{1.3\times10^{19}}$.

7. $J = \frac{I}{A}$, $\Rightarrow$ $I = JA = (300 \text{ A/m}^2)(\pi)(2.5\times10^{-3} \text{ m})^2 = 5.89\times10^{-3} \text{ A}$,

$I = \frac{\Delta q}{\Delta t}$, $\Rightarrow$ $\Delta t = \frac{\Delta q}{I} = \frac{(6.02\times10^{23})(1.60\times10^{-19} \text{ C})}{5.89\times10^{-3} \text{ A}} = 1.64\times10^7 \text{ s} \approx \boxed{190 \text{ days}}$.

13. $A = \frac{\pi d^2}{4}$, So, $\frac{A_2}{A_1} = (\frac{d_2}{d_1})^2 = 4:1$, $R = \frac{\rho L}{A}$, So, $\frac{R_1/L_1}{R_2/L_2} = \frac{A_2}{A_1} = \boxed{4:1}$.

19. (a) $R = \frac{\rho L}{A}$, $\Rightarrow$ $\rho = \frac{RA}{L} = \frac{(0.12 \text{ }\Omega)(1.0\times10^{-3} \text{ m})^2}{5.0 \text{ m}} = \boxed{2.4\times10^{-8} \text{ }\Omega\cdot\text{m}}$,

(b) $I = \frac{V}{R} = \frac{0.36 \text{ V}}{0.12 \text{ }\Omega} = \boxed{3.0 \text{ A}}$, (c) $E = \frac{V}{L} = \frac{0.36 \text{ V}}{5.0 \text{ m}} = \boxed{0.072 \text{ V/m}}$,

(d) $J = \frac{I}{A} = \frac{3.0 \text{ A}}{(1.0\times10^{-3} \text{ m})^2} = \boxed{3.0\times10^6 \text{ A/m}^2}$,

(e) $J = nev_d$, $\Rightarrow$ $v_d = \frac{J}{ne} = \frac{3.0\times10^6 \text{ A/m}^2}{(6.0\times10^{28} \text{ /m}^3)(1.60\times10^{-19} \text{ C})} = \boxed{3.1\times10^{-4} \text{ m/s}}$.

25. $R_0 = \frac{V}{I_0} = \frac{110 \text{ V}}{3.0 \text{ A}} = 36.7 \text{ }\Omega$, $R = \frac{V}{I} = \frac{110 \text{ V}}{2.7 \text{ A}} = 40.7 \text{ }\Omega$,

$R = R_0[1 + \alpha(T - T_0)]$, $\Rightarrow$ $T = T_0 + \frac{1}{\alpha}(\frac{R}{R_0} - 1) = 20°\text{C} + \frac{1}{4\times10^{-4} \text{ /C}°}(\frac{40.7 \text{ }\Omega}{36.7 \text{ }\Omega} - 1) = \boxed{290°\text{C}}$.

31. (a) $\varepsilon = \boxed{12.0 \text{ V}}$, (b) $r = \dfrac{\varepsilon}{I} = \dfrac{12.0 \text{ V}}{100 \text{ A}} = \boxed{0.12 \ \Omega}$,

(c) $\varepsilon = Ir + IR = I(r + R)$, $\Rightarrow$ $I = \dfrac{\varepsilon}{R + r} = \dfrac{12.0 \text{ V}}{0.12 \ \Omega + 5.0 \ \Omega} = \boxed{2.3 \text{ A}}$,

(d) $V = IR = (2.34 \text{ A})(5.0 \ \Omega) = \boxed{11.7 \text{ V}}$.

37. In internal resistor, $P_r = I^2 r = (2.34 \text{ A})^2 (0.12 \ \Omega) = \boxed{0.66 \text{ W}}$,

In resistor, $P_R = I^2 R = (2.34 \text{ A})^2 (5.0 \ \Omega) = \boxed{27.4 \text{ W}}$,

Output of battery, $P = \varepsilon I = (12.0 \text{ V})(2.34 \text{ A}) = \boxed{28.1 \text{ W}}$.

43. $P = IV$, $\Rightarrow$ $I = \dfrac{P}{V} = \dfrac{40 \text{ W}}{12 \text{ V}} = \boxed{3.3 \text{ A}}$.

49. $E = Pt = (IV)t = V(It) = (12 \text{ V})(300 \text{ A·h})(3600 \text{ s/h}) = \boxed{1.3 \times 10^7 \text{ J}}$.

55. $J(4\pi r^2) = I$, $\Rightarrow$ $\dfrac{E}{\rho}(4\pi r^2) = I$, So, $E = \dfrac{\rho I}{4\pi r^2}$.

$\Delta V = \displaystyle\int_{r_1}^{r_2} E\, dr = \dfrac{\rho I}{4\pi} \int_{r_1}^{r_2} \dfrac{dr}{r^2} = \dfrac{\rho I}{4\pi}\left(\dfrac{1}{r_1} - \dfrac{1}{r_2}\right)$, So, $R = \dfrac{\Delta V}{I} = \boxed{\dfrac{\rho}{4\pi}\left(\dfrac{1}{r_1} - \dfrac{1}{r_2}\right)}$.

Chapter 27 Direct Current Circuits

1.

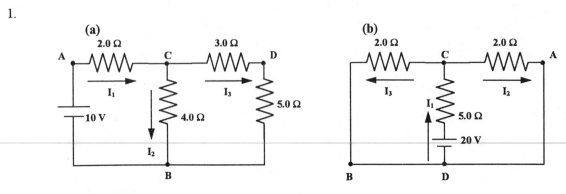

(a) For junction C , $I_1 = I_2 + I_3$,

For loop ACBA , $(2.0 \ \Omega)I_1 + (4.0 \ \Omega)I_2 = 10 \text{ V}$,

For loop CDBC , $(3.0\ \Omega)I_3 + (5.0\ \Omega)I_3 = (4.0\ \Omega)I_2$,

Solve for $I_1 = \boxed{2.1\ A}$, $I_2 = \boxed{1.4\ A}$, and $I_3 = \boxed{0.71\ A}$.

(b) For junction C , $I_1 = I_2 + I_3$,

For loop CADC , $(2.0\ \Omega)I_2 + (5.0\ \Omega)I_1 = 20\ V$,

For loop CBDC , $(2.0\ \Omega)I_3 + (5.0\ \Omega)I_1 = 20\ V$,

Solve for $I_1 = \boxed{3.3\ A}$, and $I_2 = I_3 = \boxed{1.7\ A}$.

(c)

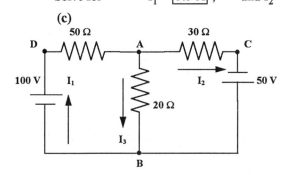

(d)

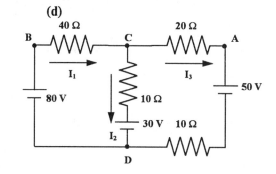

(c) For junction A , $I_1 = I_2 + I_3$,

For loop DABD , $(50\ \Omega)I_1 + (20\ \Omega)I_3 = 100\ V$,

For loop ACBA , $(30\ \Omega)I_2 - (20\ \Omega)I_3 = 50\ V$,

Solve for $I_1 = \boxed{1.9\ A}$, $I_2 = \boxed{1.8\ A}$, and $I_3 = \boxed{0.16\ A}$.

(d) For junction C , $I_1 = I_2 + I_3$,

For loop BCDB , $(40\ \Omega)I_1 + (10\ \Omega)I_2 = 80\ V - 30\ V = 50\ V$,

For loop CADC , $(20\ \Omega)I_3 + (10\ \Omega)I_3 - (10\ \Omega)I_2 = 30\ V - 50\ V = -20\ V$,

Solve for $I_1 = \boxed{0.95\ A}$, $I_2 = \boxed{1.2\ A}$, and $I_3 = \boxed{-0.26\ A}$.

The – sign means I_3 is actually opposite to shown,

(e)

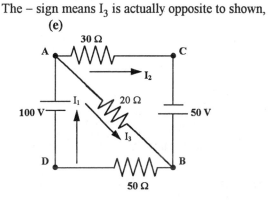

(f)

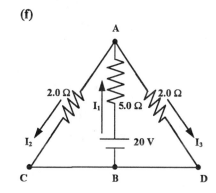

(e) For junction A , $I_1 = I_2 + I_3$,

For loop ABDA , $(20\ \Omega)I_3 + (50\ \Omega)I_1 = 100\ V$,

For loop ACBA , $(30\ \Omega)I_2 - (20\ \Omega)I_3 = 50\ V$,

Solve for $I_1 = \boxed{1.9\ A}$, $I_2 = \boxed{1.8\ A}$, and $I_3 = \boxed{0.16\ A}$.

(f) For junction A , $I_1 = I_2 + I_3$,

For loop ACBA , $(2.0 \ \Omega)I_2 + (5.0 \ \Omega)I_1 = 20 \text{ V}$,

For loop ADBA , $(2.0 \ \Omega)I_3 + (5.0 \ \Omega)I_1 = 20 \text{ V}$,

Solve for $I_1 = \boxed{3.3 \text{ A}}$, $I_2 = I_3 = \boxed{1.7 \text{ A}}$.

7. $V_A - (10 \ \Omega)I + 20 \text{ V} - (20 \ \Omega)I - 10 \text{ V} = V_B$, So, $I = \dfrac{-20 \text{ V} + 20 \text{ V} - 10 \text{ V}}{10 \ \Omega + 20 \ \Omega} = \boxed{-0.33 \text{ A}}$,

The – sign means the current is opposite to shown.

13. For junctions A and D , $I = I_1 + I_3 = I_2 + I_4$,

For junctions B and C , $I_1 = I_2 + I_5$, $I_3 + I_5 = I_4$,

For loop ABDA , $(20 \ \Omega)I_1 + (40 \ \Omega)I_2 = 20 \text{ V}$,

For loop ACDA , $(40 \ \Omega)I_3 + (20 \ \Omega)I_4 = 20 \text{ V}$,

For loop BDCB , $(40 \ \Omega)I_2 = (20 \ \Omega)I_4 + (50 \ \Omega)I_5$,

Solve for $I_1 = I_4 = \boxed{0.39 \text{ A}}$, $I_2 = I_3 =$

$\boxed{0.30 \text{ A}}$, $I_5 = \boxed{0.087 \text{ A}}$, and $I = \boxed{0.70 \text{ A}}$.

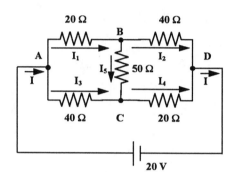

19. For (e) , $R = \boxed{15 \ \Omega}$, For (f) , the resistance between C and D is made zero by connecting C and D.

$R_s = 4 \ \Omega + 4 \ \Omega + 4 \ \Omega = 12 \ \Omega$,

$R_p = \dfrac{(12 \ \Omega)(12 \ \Omega)}{12 \ \Omega + 12 \ \Omega} = 6 \ \Omega$,

$R_s = 6 \ \Omega + 6 \ \Omega = 12 \ \Omega$,

$R_p = \dfrac{(18 \ \Omega)(12 \ \Omega)}{18 \ \Omega + 12 \ \Omega} = 7.2 \ \Omega$,

$R = 9 \ \Omega + 7.2 \ \Omega + 9 \ \Omega = \boxed{25 \ \Omega}$.

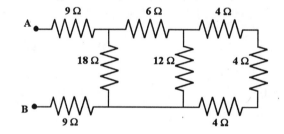

25. The two 15 Ω's on right in series, $R_s = 15 \ \Omega + 15 \ \Omega = 30 \ \Omega$,

20 Ω and 30 Ω on left in parallel , $R_p = \dfrac{(20 \ \Omega)(30 \ \Omega)}{20 \ \Omega + 30 \ \Omega} = 12 \ \Omega$,

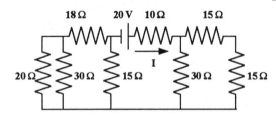

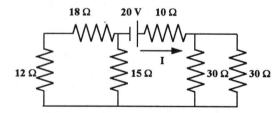

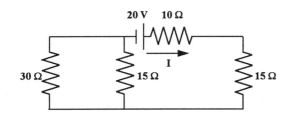

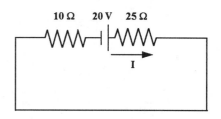

The two 30 Ω's on right in parallel, $\qquad$ $R_p = \dfrac{(30\ \Omega)(30\ \Omega)}{30\ \Omega + 30\ \Omega} = 15\ \Omega$,

The 12 Ω on left in series with 18 Ω , $\qquad$ $R_s = 12\ \Omega + 18\ \Omega = 30\ \Omega$,

The 15 Ω on right in series with 10 Ω , $\qquad$ $R_s = 15\ \Omega + 10\ \Omega = 25\ \Omega$,

The 30 Ω on left in parallel with 15 Ω , $\qquad$ $R_p = \dfrac{(30\ \Omega)(15\ \Omega)}{30\ \Omega + 15\ \Omega} = 10\ \Omega$,

Finally, the 25 Ω and 10 Ω in series, $\qquad$ $R = 25\ \Omega + 10\ \Omega = 35\ \Omega$,

(a) The current through the 10 Ω is the same as through the battery. $\quad I = \dfrac{\varepsilon}{R} = \dfrac{20\ \text{V}}{35\ \Omega} = \boxed{0.57\ \text{A}}$,

(b) The potential difference across the 15 Ω on left is $V = (0.571\ \text{A})(10\ \Omega) = 5.71\ \text{V}$,

Then, the current through the 18 Ω is $\dfrac{5.71\ \text{V}}{30\ \Omega} = 0.190\ \text{A}$,

Therefore the potential difference across the 20 Ω is $(0.190\ \text{A})(12\ \Omega) = \boxed{2.3\ \text{V}}$.

31. (a) By connecting a parallel (shunt) resistor R .

$R = \dfrac{(I_g)R_g}{I_f - (I_g)_f} = \dfrac{(0.50\times10^{-3}\ \text{A})(100\ \Omega)}{2.0\ \text{A} - 0.50\times10^{-3}\ \text{A}} = \boxed{0.025\ \Omega}$,

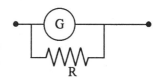

(b) $R = \dfrac{V_f}{(I_g)_f} - R_g = \dfrac{150\ \text{V}}{0.50\times10^{-3}\ \text{A}} - 100\ \Omega = \boxed{300\ \text{k}\Omega}$.

37. $\dfrac{R_x}{R_3} = \dfrac{R_1}{R_2} = \dfrac{L_1}{L_2}$, $\qquad \Rightarrow \qquad R_x = \dfrac{L_1 R_3}{L_2} = \dfrac{(27.3\ \text{cm})(25.0\ \Omega)}{72.7\ \text{cm}} = \boxed{9.39\ \Omega}$.

43. (a) $P = I^2 R = \left(\dfrac{\varepsilon}{R+r}\right)^2 R = \dfrac{\varepsilon^2 R}{(R+r)^2}$,

$\dfrac{dP}{dR} = 0 = \dfrac{\varepsilon^2}{(R+r)^2} - \dfrac{2\varepsilon^2 R}{(R+r)^3}$, $\qquad \Rightarrow \qquad R + r = 2R$,

So, $\quad R = \boxed{r}$, $\qquad$ and $\qquad P_{max} = \dfrac{\varepsilon^2 r}{(2r)^2} = \boxed{\dfrac{\varepsilon^2}{4r}}$.

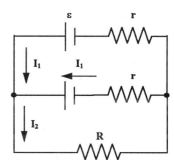

(b) Since the two sources are identical, the currents through them are the same.

For the lower loop, $\qquad I_1 r + I_2 R = \varepsilon$, $\qquad I_1 = \dfrac{I_2}{2}$, $\qquad$ So, $\qquad I_2 = \dfrac{\varepsilon}{R + r/2}$,

Power $P = I_2{}^2 R = \dfrac{\varepsilon^2 R}{(R + r/2)^2}$, $\qquad \dfrac{dP}{dR} = \dfrac{\varepsilon^2}{(R + r/2)^2} - \dfrac{2\varepsilon^2 R}{(R + r/2)^3} = \dfrac{\varepsilon^2}{(R + r/2)^2}\left[1 - \dfrac{2R}{R + r/2}\right] = 0$

Therefore, $\qquad R = \boxed{r/2}$, $\qquad\qquad\qquad P_{max} = \boxed{\dfrac{\varepsilon^2}{2r}}$.

49. $\qquad I = \dfrac{V}{R_x + R_A}$, $\quad$ or $\qquad R_x = \dfrac{V}{I} - R_A = \boxed{R'_x - R_A}$,

$\qquad R'_x = \dfrac{1.50\ V}{15.0\ mA} = 100\ \Omega$, $\qquad\qquad R_x = 100\ \Omega - 0.20\ \Omega = 99.8\ \Omega$, $\quad \Delta R = \boxed{0.2\ \Omega}$.

55. $\qquad$ For (b), when A and B are short circuited, 3.0 Ω and 6.0 Ω in parallel, $R_p = \dfrac{(3.0\ \Omega)(6.0\ \Omega)}{3.0\ \Omega + 6.0\ \Omega} = 2.0\ \Omega$,

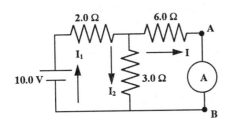 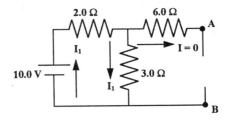

This 2.0 Ω and 2.0 Ω in series, $\qquad\qquad R_s = 2.0\ \Omega + 2.0\ \Omega = 4.0\ \Omega$,

$I_1 = \dfrac{10.0\ V}{4.0\ \Omega} = 2.5\ A$, $\qquad\qquad\qquad I = \dfrac{(2.5\ A)(2.0\ \Omega)}{6.0\ \Omega} = 0.833\ A$,

When A and B are open , $\quad I_1 = \dfrac{10.0\ V}{2.0\ \Omega + 3.0\ \Omega} = 2.0\ A$, $\qquad V = (2.0\ A)(3.0\ \Omega) = 6.0\ V$,

So, $\varepsilon_T = \boxed{6.0\ V}$, $R_T = \dfrac{\varepsilon_T}{I} = \dfrac{6.0\ V}{0.833\ A} = \boxed{7.2\ \Omega}$. The current through R is $\dfrac{6.0\ V}{7.2\ \Omega + 4.0\ \Omega} = \boxed{0.54\ A}$,

For (c), when A and B are short circuited,

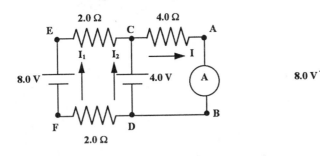 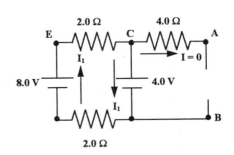

For junction C , $\qquad\qquad\qquad I = I_1 + I_2$,

For loop CABDC , $\qquad\qquad (4.0\ \Omega)I = 4.0\ V$, $\qquad$ So, $\qquad I = 1.0\ A$,

When A and B are open, $\qquad I_1 = \dfrac{8.0\ V - 4.0\ V}{2.0\ \Omega + 2.0\ \Omega} = 1.0\ A$, $\qquad V = 4.0\ V$,

So, $\qquad \varepsilon_T = \boxed{4.0\ V}$, $\qquad R_T = \dfrac{4.0\ V}{1.0\ A} = \boxed{4.0\ \Omega}$, $\quad$ The current through R is $\dfrac{4.0\ V}{4.0\ \Omega + 8.0\ \Omega} = \boxed{0.33\ A}$

Chapter 28 The Magnetic Field

1. (a) $\boxed{\text{upward}}$ since electron has negative charge. (b) $\boxed{\text{Eastward}}$.

7. For $\mathbf{v}_1 = v_0\,\mathbf{k}$, $\mathbf{F}_m = q\mathbf{v}\times\mathbf{B} = e(v_0\,\mathbf{k})\times(B_0\,\mathbf{k}) = \boxed{0}$,

   For $\mathbf{v}_2 = v_0\,\mathbf{i}$, $\mathbf{F}_m = e(v_0\,\mathbf{i})\times(B_0\,\mathbf{k}) = \boxed{-ev_0 B_0\,\mathbf{j}}$,

   For $\mathbf{v}_3 = v_0\,\mathbf{j}$, $\mathbf{F}_m = e(v_0\,\mathbf{j})\times(B_0\,\mathbf{k}) = \boxed{ev_0 B_0\,\mathbf{i}}$,

   For $\mathbf{v}_4 = v_0(\frac{1}{2}\mathbf{i} + \frac{\sqrt{3}}{2}\mathbf{j})$, $\mathbf{F}_m = e[\,v_0(\frac{1}{2}\mathbf{i} + \frac{\sqrt{3}}{2}\mathbf{j})]\times(B_0\,\mathbf{k}) = \boxed{\dfrac{ev_0 B_0}{2}(\sqrt{3}\,\mathbf{i} - \mathbf{j})}$.

13. $KE = qV = \frac{1}{2}mv^2$, $\Rightarrow$ $v = \sqrt{\dfrac{2qV}{m}}$, $R = \dfrac{mv}{qB} = \dfrac{m}{qB}\sqrt{\dfrac{2qV}{m}} = \boxed{\dfrac{1}{B}\sqrt{\dfrac{2mV}{q}}}$

19. $T = \dfrac{2.0\times10^{-3}\text{ s}}{8} = 2.5\times10^{-4}\text{ s}$, $T = \dfrac{2\pi r}{v} = \dfrac{2\pi m}{qB}$,

    So, $m = \dfrac{qBT}{2\pi} = \dfrac{(1.60\times10^{-19}\text{ C})(2.0\times10^{-2}\text{ T})(2.5\times10^{-4}\text{ s})}{2\pi} = \boxed{1.3\times10^{-25}\text{ kg}}$.

31. $F_m = IlB = F_g = mg$, $\Rightarrow$ $I = \dfrac{mg}{lB} = \dfrac{(m/l)g}{B} = \dfrac{(0.050\text{ kg/m})(9.80\text{ m/s}^2)}{2.0\text{ T}} = \boxed{0.25\text{ A to right}}$.

37. (a) $\mathcal{M} = NIA$, $\Rightarrow$ $I = \dfrac{\mathcal{M}}{NA} = \dfrac{3.0\text{ A}\cdot\text{m}^2}{200(\pi)(0.020\text{ m})^2} = \boxed{12\text{ A}}$,

    (b) $\tau_{max} = \mathcal{M}B = (3.0\text{ A}\cdot\text{m}^2)(5.0\times10^{-2}\text{ T}) = \boxed{0.15\text{ N}\cdot\text{m}}$,

    (c) $\tau = \tau_{max}\sin\theta = (0.15\text{ N}\cdot\text{m})\sin45° = \boxed{0.11\text{ N}\cdot\text{m}}$,

    (d) $U = -\mathcal{M}\cdot\mathbf{B} = -(3.0\text{ A}\cdot\text{m}^2)(5.0\times10^{-2}\text{ T})\cos45° = \boxed{-0.11\text{ J}}$.

43. (a) $v_d = \dfrac{E_H}{B} = \dfrac{1.5\times10^{-3}\text{ V/m}}{2.5\text{ T}} = \boxed{6.0\times10^{-4}\text{ m/s}}$,

    (b) $j = \dfrac{I}{A} = nev_d$, $I = nev_d A = (8.0\times10^{28}\text{ /m}^3)(1.60\times10^{-19}\text{ C})(6.0\times10^{-4}\text{ m/s})(5.0\times10^{-6}\text{ m}^2) = \boxed{38\text{ A}}$,

    (c) $\dfrac{1}{nq} = \dfrac{1}{(8.0\times10^{28}\text{ /m}^3)(-1.60\times10^{-19}\text{ C})} = \boxed{-7.8\times10^{-11}\text{ m}^3\text{/C}}$.

49. (a) $KE_{max} = \dfrac{q^2B^2R^2}{2m}$,

    So, $R = \dfrac{\sqrt{2mKE_{max}}}{qB} = \dfrac{\sqrt{2(1.67\times10^{-27}\text{ kg})(20\times10^6\text{ eV})(1.60\times10^{-19}\text{ J/eV})}}{(1.60\times10^{-19}\text{ C})(1.5\text{ T})} = \boxed{0.43\text{ m}}$,

(b) $T = \dfrac{2\pi m}{qB} = \dfrac{2\pi(1.67\times10^{-27} \text{ kg})}{(1.60\times10^{-19} \text{ C})(1.5 \text{ T})} = 4.37\times10^{-8}$ s , $\qquad f = \dfrac{1}{T} = \boxed{2.3\times10^7 \text{ Hz}}$.

55.   (a) Mass $= \rho V = \rho\pi r^2 h = (8.96 \text{ g/cm}^3)(\pi)(8.0 \text{ cm})^2(0.25 \text{ cm}) = 450$ g . So there are

$\dfrac{450 \text{ g}}{63.54 \text{ g/mol}}\times(6.02\times10^{23} \text{ atoms/mol}) = 4.26\times10^{24}$ atoms and $29(4.26\times10^{24}) = 1.24\times10^{26}$ electrons.

Therefore the net magnetic dipole moment $= (1.24\times10^{26})(9.3\times10^{-24} \text{ A·m}^2) = \boxed{1.2\times10^3 \text{ A·m}^2}$,

(b) $\mathcal{M} = IA$, $\qquad\Rightarrow\qquad I = \dfrac{\mathcal{M}}{A} = \dfrac{1.15\times10^3 \text{ A·m}^2}{\pi(0.080 \text{ m})^2} = \boxed{5.7\times10^4 \text{ A}}$.

61.   From each infinitesimal rectangular element, $\qquad d\tau = d\mathcal{M} \times \mathbf{B}$,

$$\tau = \int d\mathcal{M}\times\mathbf{B} = \left(\int d\mathcal{M}\right) \times \mathbf{B} = \mathcal{M} \times \mathbf{B} ,\qquad \text{where, because they flow in opposite directions, the}$$

imaginary currents internal to the loop cancel in pairs.

Chapter 29 Properties of the Magnetic Field

1. $B = \dfrac{\mu_0 I}{2\pi R} = \dfrac{(4\pi\times10^{-7} \text{ T·m/A})(10^4 \text{ A})}{2\pi(1.0 \text{ m})} = \boxed{2.0\times10^{-3} \text{ T}}$.

7.   (a) At P_1 , $\qquad B_1 = B_2 = \dfrac{\mu_0 I}{2\pi R} = \dfrac{\mu_0 I}{2\pi a}$ in ,

So, $\qquad B = B_1 + B_2 = \boxed{\dfrac{\mu_0 I}{\pi a} \text{ in}}$,

(b) At P_2 , $\qquad B_1 = \dfrac{\mu_0 I}{2\pi(4a)}$ in $= \dfrac{\mu_0 I}{8\pi a}$ in ,

$B_2 = \dfrac{\mu_0 I}{2\pi(2a)}$ out $= \dfrac{\mu_0 I}{4\pi a}$ out ,

So, $\qquad B = B_2 - B_1 = \boxed{\dfrac{\mu_0 I}{8\pi a} \text{ out}}$.

13.   At center, $\qquad B_0 = \dfrac{\mu_0 I}{2R}$, $\qquad$ At y from center , $\qquad B = \dfrac{\mu_0 I R^2}{2(y^2 + R^2)^{3/2}}$,

$B = B_0/2$, $\qquad$ So, $\qquad \dfrac{R^2}{(y^2 + R^2)^{3/2}} = \dfrac{1}{2R}$, $\qquad$ Solve for $\qquad y = \boxed{0.77R}$.

18. Due to symmetry, the magnetic field at P of the two semicircles (the two small horizontal segments have zero contributions since R = 0) is half of the magnetic field by two full circles.   For a full circle, the magnetic field at the center is $B = \frac{\mu_0 I}{2R}$,   So, $B = \frac{\mu_0 I}{4R}$ is for semicircle.

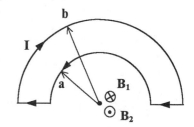

$B_1 = \frac{\mu_0 I}{4b}$ in , $B_2 = \frac{\mu_0 I}{4a}$ out ,

$B = B_2 - B_1 = \frac{\mu_0 I}{4} (\frac{1}{a} - \frac{1}{b})$ out $= \boxed{\frac{\mu_0 I}{4 ab} (b - a) \text{ out}}$.

19. Due to symmetry, the magnetic field at P is half of the result in Problem 29-18 . $B = \boxed{\frac{\mu_0 I}{8ab} (b - a) \text{ out}}$

25. $\boxed{\text{Impossible}}$ since $\oint \mathbf{B} \cdot \mathbf{n} da \neq 0$ for this $\mathbf{B}$.

31. Along the front face,   x = 1.0 m ,   So, $\mathbf{B} = b_1 \mathbf{i} + b_2 y \mathbf{j}$,

$$\oint \mathbf{B} \cdot dl = \int_0^{1.0\text{ m}} (b_1 \mathbf{i} + b_2 y \mathbf{j}) \cdot (\mathbf{j}) dy + \int_{1.0\text{ m}}^{0} (b_1 \mathbf{i} + b_2 y \mathbf{j}) \cdot (\mathbf{k}) dz + \int_{1.0\text{ m}}^{0} (b_1 \mathbf{i} + b_2 y \mathbf{j}) \cdot (\mathbf{j}) dy$$

$$+ \int_0^{1.0\text{ m}} (b_1 \mathbf{i} + b_2 y \mathbf{j}) \cdot (\mathbf{k}) dz = 0 = \mu_0 I_s ,   \text{So,}   I_s = \boxed{0} .$$

37.

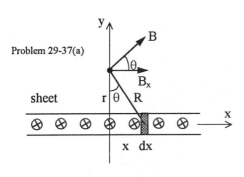 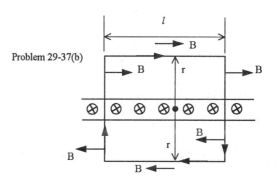

(a) From an infinitely long and infinitesimal strip of current of width dx (with current jdx), the magnetic field at a point r above the sheet is $dB = \frac{\mu_0 dI}{2\pi R} = \frac{\mu_0 jdx}{2\pi \sqrt{r^2 + x^2}}$,

Due to symmetry, the magnetic field is in the $\boxed{+x}$ axis. The y components of the magnetic fields by all infinitesimal strips cancel out.

$$dB_x = dB\cos\theta = \frac{\mu_0 jdx}{2\pi\sqrt{r^2 + x^2}} \cdot \frac{r}{\sqrt{r^2 + x^2}} = \frac{\mu_0 jrdx}{2\pi(r^2 + x^2)} ,$$

$$B_x = \int_{-\infty}^{\infty} dB_x = \frac{\mu_0 jr}{2\pi} \int_{-\infty}^{\infty} \frac{dx}{r^2 + x^2} = \frac{\mu_0 jr}{2\pi} \cdot [\frac{1}{r}\tan^{-1}(\frac{x}{r})]\Big|_{-\infty}^{\infty} = \frac{\mu_0 jr}{2\pi} \cdot (\frac{\pi}{r}) = \boxed{\frac{\mu_0 j}{2}} ,$$

(b) $\oint \mathbf{B} \cdot d l = \mu_0 I_s ,$ The left and right segments have zero contribution since $\mathbf{B} \cdot d l = 0$,

For the upper and lower segments, $\mathbf{B} \cdot d l = Bd l ,$ So $\oint \mathbf{B} \cdot d l = 2B l = \mu_0 I_s = \mu_0(j l) ,$ $B = \boxed{\frac{\mu_0 j}{2}}$.

43.

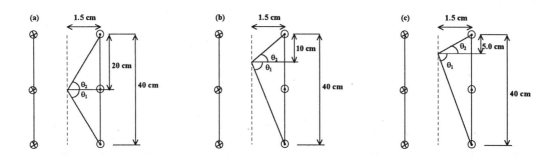

(a) $\theta_1 = -\tan^{-1}(\frac{20}{1.5}) = -85.7° ,$ $\theta_2 = 85.7° ,$

$B = \frac{\mu_0 nI}{2}(\sin\theta_2 - \sin\theta_1) = \frac{(4\pi\times10^{-7}\ \text{T·m/A})(500/0.40\ \text{m})(4.0\ \text{A})}{2}(2\sin85.7°) = \boxed{6.3\times10^{-3}\ \text{T}} ,$

(b) $\theta_1 = -\tan^{-1}(\frac{30}{1.5}) = -87.1° ,$ $\theta_2 = \tan^{-1}(\frac{10}{1.5}) = 81.5° ,$

$B = \frac{(4\pi\times10^{-7}\ \text{T·m/A})(500/0.40\ \text{m})(4.0\ \text{A})}{2}[\sin81.5° - \sin(-87.1°)] = \boxed{6.2\times10^{-3}\ \text{T}} ,$

(c) $\theta_1 = -\tan^{-1}(\frac{35}{1.5}) = -87.5° ,$ $\theta_2 = \tan^{-1}(\frac{5.0}{1.5}) = 73.3° ,$

$B = \frac{(4\pi\times10^{-7}\ \text{T·m/A})(500/0.40\ \text{m})(4.0\ \text{A})}{2}[\sin73.3° - \sin(-87.5°)] = \boxed{6.1\times10^{-3}\ \text{T}} ,$

(d) $B = \mu_0 nI = (4\pi\times10^{-7}\ \text{T·m/A})(500/0.40\ \text{m})(4.0\ \text{A}) = \boxed{6.3\times10^{-3}\ \text{T}}$.

49. $B = \frac{\mu_0 NI}{2\pi r} ,$ $dB = -\frac{\mu_0 NIdr}{2\pi r^2} = -\frac{Bdr}{r} ,$ So, $\frac{dB}{B} = \boxed{-\frac{dr}{r}} .$

55.   (a) $\mathcal{M} = (10^{19})(2.1\times10^{-23}\ \text{A·m}^2) = \boxed{2.1\times10^{-4}\ \text{A·m}^2} ,$

(b) $\mathcal{M} = IA = I(\pi r^2) ,$ $\Rightarrow$ $I = \frac{\mathcal{M}}{\pi r^2} = \frac{2.1\times10^{-4}\ \text{A·m}^2}{\pi(0.005\ \text{m})^2} = \boxed{2.7\ \text{A}} .$

61. Due to symmetry, all four segments have equal contribution to the magnetic field at P , which is perpendicular to the plane of the square.

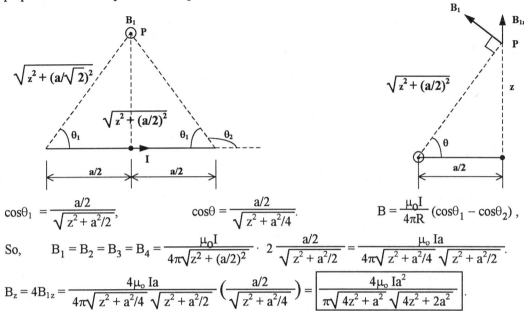

$$\cos\theta_1 = \frac{a/2}{\sqrt{z^2 + a^2/2}}, \qquad \cos\theta = \frac{a/2}{\sqrt{z^2 + a^2/4}}. \qquad B = \frac{\mu_0 I}{4\pi R}(\cos\theta_1 - \cos\theta_2) ,$$

So, $B_1 = B_2 = B_3 = B_4 = \frac{\mu_0 I}{4\pi\sqrt{z^2 + (a/2)^2}} \cdot 2 \frac{a/2}{\sqrt{z^2 + a^2/2}} = \frac{\mu_0 Ia}{4\pi\sqrt{z^2 + a^2/4}\ \sqrt{z^2 + a^2/2}}.$

$$B_z = 4B_{1z} = \frac{4\mu_0 Ia}{4\pi\sqrt{z^2 + a^2/4}\ \sqrt{z^2 + a^2/2}}\left(\frac{a/2}{\sqrt{z^2 + a^2/4}}\right) = \boxed{\frac{4\mu_0 Ia^2}{\pi\sqrt{4z^2 + a^2}\ \sqrt{4z^2 + 2a^2}}}.$$

67. From an infinitesimal long strip of width dx' (with current $\frac{Idx'}{a}$), the magnetic field at P is

$$dB = \frac{\mu_0 dI}{2\pi(x - x')} = \frac{\mu_0 I}{2\pi a} \cdot \frac{dx'}{x - x'} ,$$

$$B = \int_{-a}^{0} dB = \frac{\mu_0 I}{2\pi a}\int_{-a}^{0}\frac{dx'}{x - x'} = \frac{\mu_0 I}{2\pi a}\ln(x - x')\Big|_{-a}^{0} = -\frac{\mu_0 I}{2\pi a}\ln\frac{x}{x + a} = \frac{\mu_0 I}{2\pi a}\ln\frac{x + a}{x} = \frac{\mu_0 I}{2\pi a}\ln\left(1 + \frac{a}{x}\right).$$

When $a \to 0$, $\lim_{a\to 0}\ln\left(1 + \frac{a}{x}\right) \to \frac{a}{x},$ Therefore, $B = \boxed{\frac{\mu_0 I}{2\pi x}}.$

Chapter 30 Time-varying Magnetic Field and Faraday's Law

1. (a) $\boxed{\text{CCW viewed from the magnet}}$. (b) $\boxed{\text{CW viewed from the magnet}}$.

(c) $\boxed{\text{CW viewed from the magnet}}$. (d) $\boxed{\text{CCW viewed from the magnet}}$.

(e) $\boxed{\text{CW viewed from the magnet}}$. (f) $\boxed{\text{No current}}$.

7. $B = 1.5\,t\,,$ $0 \le t < 2.0$ ms,

$B = 3.0$ mT, 2.0 ms $\le t \le 5.0$ ms,

$B = -3.0t + 18$ mT, 5.0 ms $< t \le 6.0$ ms,

$$\varepsilon = -\frac{d\Phi_m}{dt} = -\frac{d(BA)}{dt} = -A\frac{dB}{dt}\,,$$

$\varepsilon = -\pi(0.100 \text{ m})^2(1.5 \text{ T/s})$

$= \boxed{-47 \text{ mV} \qquad (0 \le t < 2.0 \text{ ms})}\,,$

$\varepsilon = -\pi(0.100 \text{ m})^2(0) = \boxed{0 \quad (2.0 \text{ ms} \le t \le 5.0 \text{ ms})}\,,$

$\varepsilon = -\pi(0.100 \text{ m})^2(-3.0 \text{ T/s}) = \boxed{94 \text{ mV} \qquad (5.0 \text{ ms} < t \le 6.0 \text{ ms})}$.

13. $\mathbf{n} = \mathbf{k}\,,$ $d\Phi_m = \mathbf{B} \cdot \mathbf{n}\,da = Cy\sin(\omega t)dxdy\,,$

$$\Phi_m = C\sin(\omega t)\int_0^b ay\,dy = \frac{Cab^2\sin(\omega t)}{2}\,, \qquad \varepsilon = -\frac{d\Phi_m}{dt} = \boxed{-\frac{Cab^2\omega\cos(\omega t)}{2}}\,.$$

18. Inside, $B = \mu_0 nI\,,$ $\oint \mathbf{E} \cdot d\boldsymbol{l} = E(2\pi r) = \left|\frac{d\Phi_m}{dt}\right| = \left|A\frac{dB}{dt}\right| = (\pi r^2)\mu_0 n\frac{dI}{dt}\,,$

So, $E = \boxed{\dfrac{\mu_0 nr}{2} \cdot \dfrac{dI}{dt} \quad \text{(inside)}}$.

Outside, $E(2\pi r) = \pi R^2 \mu_0 n\dfrac{dI}{dt}\,,$ So, $E = \boxed{\dfrac{\mu_0 nR^2}{2r} \cdot \dfrac{dI}{dt} \quad \text{(outside)}}$

19. From Problem 30-18, $E_{\text{inside}} = \dfrac{\mu_0 nr}{2} \cdot \dfrac{dI}{dt} = \dfrac{\mu_0 nr}{2} \cdot \dfrac{d}{dt}(I_0\sin\omega t) = \boxed{\dfrac{\mu_0 nr\omega I_0\cos\omega t}{2} \quad \text{(inside)}}$,

$E_{\text{outside}} = \dfrac{\mu_0 nR^2}{2r} \cdot \dfrac{dI}{dt} = \boxed{\dfrac{\mu_0 nR^2\omega I_0\cos\omega t}{2r} \quad \text{(outside)}}$.

25. $\varepsilon = -N\dfrac{\Delta\Phi_m}{\Delta t} = -N\dfrac{\Delta(BA\cos\theta)}{\Delta t} = -NBA\dfrac{\cos 90° - \cos 0°}{0.010 \text{ s}}$

$= -(1000)(6.0\times10^{-5} \text{ T})(25\times10^{-4} \text{ m}^2)\times\dfrac{\cos 90° - \cos 0°}{0.010 \text{ s}} = \boxed{15 \text{ mV}}$.

Page 59

31. (a) $\varepsilon = Blv = (0.25\text{ T})(0.040\text{ m})(5.0\text{ m/s}) = \boxed{0.050\text{ V}}$, (b) $I = \dfrac{\varepsilon}{R} = \dfrac{0.050\text{ V}}{2.0\ \Omega} = \boxed{25\text{ mA CCW}}$,

(c) Power dissipated $= I\varepsilon = (25\text{ mA})(0.050\text{ V}) = 1.25\text{ mW} = Pv$, So, $P = \dfrac{1.25\text{ mW}}{5.0\text{ m/s}} = \boxed{2.5\times10^{-4}\text{ N}}$,

(d) As in (c), both powers are $\boxed{1.3\text{ mW}}$.

37. (a) $\omega = 3600\text{ rev/min} = (3600\text{ rev/min})(2\pi\text{ rad/rev})(1\text{ min/60 s}) = 120\pi\text{ rad/s}$,

$\varepsilon = \varepsilon_0\sin\omega t = NBA\omega\sin\omega t = (50)(0.75\text{ T})(0.15\text{ m})(0.40\text{ m})(120\pi\text{ rad/s})\sin120\pi t = \boxed{850\sin120\pi t\text{ V}}$,

(b) $P = \dfrac{\varepsilon^2}{R} = \dfrac{(848\sin120\pi t\text{ V})^2}{1000\ \Omega} = \boxed{720\sin^2120\pi t\text{ W}}$,

(c) $P = \dfrac{(848\sin120\pi t\text{ V})^2}{2000\ \Omega} = \boxed{360\sin^2120\pi t\text{ W}}$.

43. (a) $R_f + R_a = \dfrac{120\text{ V}}{2.0\text{ A}} = 60\ \Omega$, So, $R_f = 60\ \Omega - 10\ \Omega = \boxed{50\ \Omega}$,

(b) $I = \dfrac{\varepsilon_s - \varepsilon_i}{R_f + R_a}$, $\Rightarrow$ $\varepsilon_i = \varepsilon_s - I(R_f + R_a) = 120\text{ V} - (0.50\text{ A})(60\ \Omega) = \boxed{90\text{ V}}$,

(c) $\varepsilon_i = 120\text{ V} - (1.0\text{ A})(60\ \Omega) = \boxed{60\text{ V}}$.

49. $\Phi = \dfrac{\mu_o I_o a}{2\pi}\displaystyle\int_{x}^{x+b}\dfrac{dr}{r} = \dfrac{\mu_o I_o a}{2\pi}\ln\left(1 + \dfrac{b}{x}\right)$, $\varepsilon = -\dfrac{d\Phi}{dt} = -\dfrac{\mu_o I_o a}{2\pi}\dfrac{1}{1 + b/x}\dfrac{d}{dt}\left(\dfrac{b}{x}\right)$

$= -\dfrac{\mu_o I_o a}{2\pi}\dfrac{1}{1 + b/x}\left(-\dfrac{b}{x^2}\right)\dfrac{dx}{dt} = \dfrac{\mu_o I_o abv}{2\pi x(x + b)}$, So, $I = \dfrac{\varepsilon}{R} = \boxed{\dfrac{\mu_o I_o abv}{2\pi Rx(x + b)}}$.

55. (a) $B = \mu_0 nI$, $\Phi_m = BA = \mu_0 nIA$,

$\varepsilon = \dfrac{d\Phi_m}{dt} = \mu_0 nA\dfrac{dI}{dt} = (4\pi\times10^{-7}\text{ T·m/A})(10\times10^2\text{ /m})(\pi)(0.050\text{ m})^2(100\text{ A/s}) = \boxed{9.9\times10^{-4}\text{ V}}$,

(b) $\boxed{9.9\times10^{-4}\text{ V}}$,

(c) $\displaystyle\oint \mathbf{E}\cdot d\mathbf{l} = E(2\pi r) = \dfrac{d\Phi}{dt} = \varepsilon$, $\Rightarrow$ $E = \dfrac{9.87\times10^{-4}\text{ V}}{2\pi(0.100\text{ m})} = \boxed{1.6\times10^{-3}\text{ V/m}}$,

(d) $\boxed{9.9\times10^{-4}\text{ V}}$,

(e) $\boxed{\text{No because there is no cylindrical symmetry}}$.

Chapter 31 Maxwell's Equations and Electromagnetic Energy

1. (a) $B = \dfrac{\mu_o\, rI}{2\pi R^2}$, $\quad (r < R)$.

   $B = \dfrac{(4\pi\times10^{-7}\ \text{T}\cdot\text{m/A})(10^{-2}\ \text{m})(2.0\times10^{-2}\ \text{A})}{2\pi(5\times10^{-2}\ \text{m})^2}$

   $= 1.6\times10^{-8}$ T.

   (b) $B = \dfrac{(4\pi\times10^{-7}\ \text{T}\cdot\text{m/A})(2\times10^{-2}\ \text{m})(2.0\times10^{-2}\ \text{A})}{2\pi(5\times10^{-2}\ \text{m})^2}$

   $= 3.2\times10^{-8}$ T.

   (c) $B = \dfrac{\mu_o\, I}{2\pi r}$, $\quad (r > R)$.
   $\qquad B = \dfrac{(4\pi\times10^{-7}\ \text{T}\cdot\text{m/A})(2.0\times10^{-2}\ \text{A})}{2\pi(0.20\ \text{m})} = 2.0\times10^{-8}$ T.

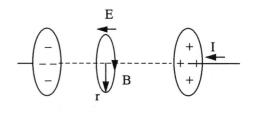

7. (a) $I_d = \varepsilon_o A\, \dfrac{dE}{dt} = \dfrac{\varepsilon_o A}{d}\, \dfrac{dV}{dt} = \dfrac{\varepsilon_o A}{d}\, \dfrac{d}{dt}\Big[V_o\,(1 - e^{-\alpha t})\Big] = \dfrac{\varepsilon_o A}{d}\,\alpha V_o\, e^{-\alpha t}$

   (b) $I_d = \dfrac{(8.85\times10^{-12}\ \text{C}^2/\text{N}\cdot\text{m}^2)(100\times10^{-4}\ \text{m}^2)}{2\times10^{-3}\ \text{m}}(0.25\ /\text{s})(200\ \text{V})\, e^{-0.25} = 1.7\times10^{-9}$ A.

13. $u_e = \dfrac{1}{2}\varepsilon_o E^2,\qquad E = \sqrt{\dfrac{2u_e}{\varepsilon_o}} = \sqrt{\dfrac{2(18\ \text{J/m}^3)}{8.85\times10^{-12}\ \text{C}^2/\text{N}\cdot\text{m}^2}} = 2.0\times10^6$ V/m.

18. $U_e = \dfrac{1}{2}\varepsilon_o E^2(\text{Volume}),\qquad E = \dfrac{Q}{A\varepsilon_o},\qquad \text{Volume} = Ad.$

   $U_e = \dfrac{1}{2}\varepsilon_o\Big(\dfrac{Q}{A\varepsilon_o}\Big)^2(Ad) = \dfrac{1}{2}\dfrac{Q^2 d}{A\varepsilon_o} = \dfrac{1}{2}\dfrac{(2.0\times10^{-7}\ \text{C})^2(2.0\times10^{-3}\ \text{m})}{(10^{-2}\ \text{m}^2)(8.85\times10^{-12}\ \text{C}^2/\text{N}\cdot\text{m}^2)} = 4.52\times10^{-4}$ J.

19. (a) From Problem 31-18, $U_e = \dfrac{1}{2}\dfrac{Q^2 d}{A\varepsilon_o}$.

   (b) Now $\qquad d \quad\Rightarrow\quad 2d,\qquad\qquad U_e = \dfrac{1}{2}\dfrac{Q^2(2d)}{A\varepsilon_o} = \dfrac{Q^2 d}{A\varepsilon_o}\qquad$ Twice that of (a).

   (c) Work performed by agent who pulled plates apart.

25. Toroid $\qquad B = \dfrac{\mu_o\, NI}{2\pi r}\qquad\qquad a \le r \le b.$

   $U_m = \displaystyle\int_a^b \dfrac{B^2}{2\mu_o}\, dV = \int_a^b \dfrac{1}{2\mu_o}\Big(\dfrac{\mu_o\, NI}{2\pi r}\Big)^2 (h)(2\pi r\, dr) = \int_a^b \dfrac{\mu_o\, N^2 I^2 h}{4\pi}\, \dfrac{dr}{r} = \dfrac{\mu_o\, N^2 I^2 h}{4\pi}\ln\Big(\dfrac{b}{a}\Big).$

31.　(a) $B = \dfrac{\mu_o I}{2\pi r} = \dfrac{(4\pi \times 10^{-7}\ T\cdot m/A)(40\ A)}{2\pi(10^{-3}\ m)} = 8.0 \times 10^{-3}\ T.$

(b) $\dfrac{R}{L} = \dfrac{\rho}{A} = \dfrac{\rho}{\pi r^2} = \dfrac{2.65 \times 10^{-8}\ \Omega\cdot m}{\pi(10^{-3}\ m)^2} = 8.44 \times 10^{-3}\ \Omega/m.$

$E = \dfrac{V}{L} = I\dfrac{R}{L} = (40\ A)(8.44 \times 10^{-3}\ \Omega/m) = 0.337\ V/m.$

(c) $S = \dfrac{E \times B}{\mu_o} = \dfrac{(0.337\ V/m)(8.0 \times 10^{-3}\ T)}{4\pi \times 10^{-7}\ T\cdot m/A}$

$= 2.15 \times 10^3\ W/m^2$　　　　(radially in).

(d) $\dfrac{Power}{meter} = S(2\pi r) = (2.15 \times 10^{-3}\ W/m^2)(2\pi)(10^{-3}\ m)(1.0\ m) = 13.5\ W.$

(e) $\dfrac{Power}{meter} = I^2 R = (40\ A)^2(8.44 \times 10^{-3}\ \Omega/m)(1.0\ m) = 13.5\ W.$

36.　$r < R,$　　$\oint \mathbf{E} \cdot \mathbf{n}\ da = \dfrac{1}{\varepsilon_o} \int \rho dV,$　　$E(4\pi r^2) = \dfrac{\rho_o}{\varepsilon_o}\dfrac{4}{3}\pi r^3,$　　$E(r) = \dfrac{\rho_o r}{3\varepsilon_o}.$

$r > R,$　　$E(r) = \dfrac{q}{4\pi\varepsilon_o r^2},$

$U'_e = \int_0^R \dfrac{\varepsilon_o}{2} E^2(r) dV = \dfrac{\varepsilon_o}{2}\int_0^R \left(\dfrac{\rho_o r}{3\varepsilon_o}\right)^2 4\pi r^2 dr = \dfrac{2\pi\rho_o{}^2 R^5}{45\varepsilon_o} = \dfrac{1}{8\pi\varepsilon_o}\left(\dfrac{q^2}{5R}\right).$

(b) $U''_e = \dfrac{1}{2}\varepsilon_o\int_R^\infty \left(\dfrac{q}{4\pi\varepsilon_o r^2}\right)^2 4\pi r^2 dr = \dfrac{q^2}{8\pi\varepsilon_o}\dfrac{1}{R},$　　$U = U'_e + U''_e = \dfrac{1}{4\pi\varepsilon_o}\dfrac{3q^2}{5R}.$

37.　From Problem 31-36,　　$E = \dfrac{\rho r}{3\varepsilon_o},$　　$q = \dfrac{4}{3}\pi R^3 \rho.$

$U_e = \int_0^R \dfrac{\varepsilon_o}{2}\left(\dfrac{\rho r}{3\varepsilon_o}\right)^2 4\pi r^2 dr = \dfrac{2\pi\rho^2 R^5}{45\varepsilon_o} = \dfrac{1}{8\pi\varepsilon_o}\dfrac{q^2}{5R} = \dfrac{1}{8\pi(8.85 \times 10^{-12}\ C^2/N\cdot m^2)}\dfrac{(1.60 \times 10^{-19}\ C)^2}{5(1.0 \times 10^{-15}\ m)}$

$= 2.30 \times 10^{-14}\ J = 1.44 \times 10^5\ eV.$

Chapter 32 Capacitance and Inductance

1.　$A = \dfrac{dC}{\varepsilon_o} = \dfrac{(2.0 \times 10^{-3}\ m)(5.0 \times 10^{-12}\ F)}{8.85 \times 10^{-12}\ C^2/N\cdot m^2} = 1.1 \times 10^{-3}\ m^2.$

7. $C = 4\pi\varepsilon_o R = 4\pi(8.85\times10^{-12}\ C^2/N{\cdot}m^2)(6.40\times10^6\ m) = 7.12\times10^{-4}\ F = 712\ \mu F.$

13. (a) $C = \dfrac{Q}{V} = \dfrac{0.020\times10^{-6}\ F}{250\ V} = 8.0\times10^{-11}\ F.$

    (b) $A = \dfrac{dC}{\varepsilon_o} = \dfrac{(4.0\times10^{-4}\ m)(8.0\times10^{-11}\ F)}{8.85\times10^{-12}\ C^2/N{\cdot}m^2} = 3.6\times10^{-3}\ m^2 = 36\ cm^2.$

    (c) $Q = CV = (8.0\times10^{-11}\ F)(500\ V) = 4.0\times10^{-8}\ C.$ (d)$V_{max} = E_{max}d = (3.0\times10^6\ V/m)(4.0\times10^{-4}\ m) = 1200\ V.$

19. Parallel, $C_p = C_1 + C_2 + C_3 = 3.0\ \mu F.$ Series, $\dfrac{1}{C_s} = \dfrac{1}{C_1} + \dfrac{1}{C_2} + \dfrac{1}{C_3} = \dfrac{3}{1.0\ \mu F},\ C_s = 0.33\ \mu F.$

25. (a) Left side, $\dfrac{1}{C_s} = \dfrac{1}{2.0\ \mu F} + \dfrac{1}{4.0\ \mu F} = 7.5\times10^5\ /F,$ $C_s = 1.33\ \mu F.$

    $Q = C_s V = (1.33\times10^{-6}\ F)(300\ V) = 4.0\times10^{-4}\ C.$ $V_D = \dfrac{Q}{C} = \dfrac{4.0\times10^{-4}\ C}{4.0\times10^{-6}\ F} = 100\ V.$

    Right side, $\dfrac{1}{C_s} = \dfrac{1}{4.0\ \mu F} + \dfrac{1}{2.0\ \mu F} = 7.5\times10^5\ /F,$ $C_s = 1.33\ \mu F.$

    $Q = C_s V = (1.33\times10^{-6}\ F)(300\ V) = 4.0\times10^{-4}\ C.$ $V_E = \dfrac{Q}{C} = \dfrac{4.0\times10^{-4}\ C}{2.0\times10^{-6}\ F} = 200\ V.$

    $V_E - V_D = 100\ V.$

    (b) Circuit becomes:

    $V_D = 150\ V.$

    (c) For $4\ \mu F,$

    $Q_4 = CV = (4.0\times10^{-6}\ F)(150\ V) = 6.0\times10^{-4}\ C.$ Previous charge $= 4.0\times10^{-4}\ C,$ $\Delta Q = 2.0\times10^{-4}\ C,$

    For $2\ \mu F,$ $Q_2 = (2.0\times10^{-6}\ F)(150\ V) = 3.0\times10^{-4}\ C.$ Previous charge $= 4.0\times10^{-4}\ C.$

    Inside plate of $4.0\ \mu F$ attached to E gained $2\times10^{-4}\ C$ of negative charge and inside plate of $2.0\ \mu F$ gained $1.0\times10^{-4}\ C$ of negative charge so $3.0\times10^{-4}\ C$ of negative charge flowed across switch from D to E.

31. $U = \dfrac{1}{2}CV^2 = \dfrac{1}{2}(8.0\times10^{-6}\ C)(6.0\ V)^2 = 1.44\times10^{-4}\ J.$

37. (a) $C = 5.53\times10^{-10}\ F,$ $V = 50\ V.$ $U_e = \dfrac{1}{2}CV^2 = \dfrac{1}{2}(5.53\times10^{-10}\ F)(50\ V)^2 = 6.9\times10^{-7}\ J.$

    (b) $C = 2.77\times10^{-10}\ F,$ $V = 50\ V.$ $U_e = \dfrac{1}{2}(2.77\times10^{-10}\ F)(50\ V)^2 = 3.5\times10^{-7}\ J.$

    Work is done by energy stored in capacitors on the battery as charge is returned to the battery.

43. (a)

$$\frac{1}{C_s} = \frac{1}{3 \times 10^{-6} \text{ F}} + \frac{1}{9 \times 10^{-6} \text{ F}} + \frac{1}{6 \times 10^{-6} \text{ F}} = 6.11 \times 10^5 \text{ /F}, \qquad C_s = 1.64 \times 10^{-6} \text{ F}.$$

$$U_e = \frac{1}{2} C_s V^2 = \frac{1}{2}(1.64 \times 10^{-6} \text{ F})(400 \text{ V})^2 = 0.131 \text{ J}.$$

(b) No, because energy is lost to resistive heating in the wires (See Chapter 33).

49. (a) $C = \dfrac{\kappa \varepsilon_o A}{d} = \dfrac{(2.1)(8.85 \times 10^{-12} \text{ C}^2/\text{N·m}^2)(50 \times 10^{-4} \text{ m}^2)}{0.50 \times 10^{-3} \text{ m}} = 1.86 \times 10^{-10} \text{ F}.$

$Q = CV = (1.86 \times 10^{-10} \text{ F})(200 \text{ V}) = 3.7 \times 10^{-8} \text{ C}.$

(b) $E = \dfrac{E_o}{\kappa} = \dfrac{V}{d} = \dfrac{200 \text{ V}}{0.5 \times 10^{-3} \text{ m}} = 4.0 \times 10^5 \text{ V/m}.$

(c) $Q_i = Q - A\varepsilon_o E = 3.717 \times 10^{-8} \text{ C} - (50 \times 10^{-4} \text{ m}^2)(8.85 \times 10^{-12} \text{ C}^2/\text{N·m}^2)(4.00 \times 10^5 \text{ V/m}) = 1.9 \times 10^{-8} \text{ C}.$

55. $\varepsilon = -L \dfrac{dI}{dt} = -(2.0 \text{ H})(240\pi)\cos(120\pi t) = -480\pi\cos(120\pi t) \text{ V} = 480\pi\sin(120\pi t - \pi/2) \text{ V}.$

61. $\dfrac{L}{l} = \dfrac{\mu_o}{2\pi}\ln\!\left(\dfrac{R_2}{R_1}\right) = \dfrac{(4\pi \times 10^{-7} \text{ T·m/A})\ln(4/0.5)}{2\pi} = 4.16 \times 10^{-7} \text{ H/m}.$

67. $U_m = \dfrac{\mu_o I^2 l}{4\pi}\ln\!\left(\dfrac{R_2}{R_1}\right) = \dfrac{(4\pi \times 10^{-7} \text{ T·m/A})(1.2 \text{ A})^2(3 \text{ m})\ln(5)}{4\pi} = 7.0 \times 10^{-7} \text{ J}.$

73. $M_{21} = \dfrac{N_2 \Phi_{21}}{I_1} = \dfrac{N_2 \mu_o N_1 A_1}{2\pi R_1} = \dfrac{(750)(4\pi \times 10^{-7} \text{ T·m/A})(1000)(0.25 \times 10^{-4} \text{ m}^2)}{2\pi(0.16 \text{ m})} = 2.3 \times 10^{-5} \text{ H}.$

79. Outside, $\qquad B = \dfrac{\mu_o I}{2\pi r},$ $\qquad\qquad$ Inside, $\quad B = \dfrac{\mu_o I r}{2\pi a^2}.$

$$U = \int \frac{B^2}{2\mu_o}dv = \frac{\mu_o^2 I^2}{8\pi^2 a^4 \mu_o}\int_0^a r^2(2\pi r l)dr + \frac{\mu_o^2 I^2}{8\pi^2 \mu_o}\int_a^R \frac{2\pi r l\,dr}{r^2} = \frac{\mu_o I^2 l}{4\pi}\left(\frac{1}{4} + \ln\frac{R}{a}\right),$$

So, $\qquad \dfrac{2U}{I^2} = \dfrac{\mu_o l}{2\pi}\left(\dfrac{1}{4} + \ln\dfrac{R}{a}\right).$ $\qquad$ and, $\qquad L = \lim_{R \to \infty}\dfrac{2U}{I^2} = \infty.$

Chapter 33 Capacitors and Inductors in Circuits

1. (a) $I_o = \dfrac{\varepsilon}{R} = \dfrac{12 \text{ V}}{5.0 \times 10^3 \text{ }\Omega} = 2.4 \times 10^{-3} \text{ A}.$ $\qquad\qquad$ (b) $\tau_c = RC = (5 \times 10^3 \text{ }\Omega)(2.0 \times 10^{-6} \text{ F}) = 0.010 \text{ s}.$

$I(0.02) = I_o e^{-t/RC} = (2.4 \times 10^{-3} \text{ A})e^{-0.020/0.010} = 3.2 \times 10^{-4} \text{ A}.$

(c) $I(\infty) = 0$.   (d) $\dfrac{dE}{dt} = I^2 R = I_o^2 \, e^{-2t/RC} R = (2.4 \times 10^{-3} \text{ A})^2 \, e^{-4.0}(5 \times 10^3 \ \Omega) = 5.3 \times 10^{-4} \text{ W}$.

(e) $U_C = \dfrac{1}{2} C V_C^2 = \dfrac{1}{2} C \varepsilon^2 \left(1 - e^{-t/\tau c}\right)^2 = \dfrac{1}{2}(2.0 \times 10^{-6} \text{ F})(12 \text{ V})^2 \left(1 - e^{-0.02/0.01}\right)^2 = 1.1 \times 10^{-4} \text{ J}$.

(f) $\dfrac{dU_C}{dt} = C V_C \dfrac{dV_C}{dt} = C\varepsilon\left(1 - e^{-t/\tau c}\right)\dfrac{\varepsilon}{\tau_c} e^{-t/\tau c} = \dfrac{\varepsilon^2}{R} e^{-t/\tau c}\left(1 - e^{-t/\tau c}\right) = \dfrac{(12 \text{ V})^2}{5 \times 10^3 \ \Omega} e^{-2}(1 - e^{-2})$

$= 3.4 \times 10^{-3} \text{ W}$.

7. $V = V_o \, e^{-t/\tau c}$, $\tau_c = \dfrac{-t}{\ln(V/V_o)} = \dfrac{-4.0 \text{ s}}{\ln(5/12)} = 4.57 \text{ s} = RC$.

$C = \dfrac{\tau_c}{R} = \dfrac{4.57 \text{ s}}{2.2 \times 10^6 \ \Omega} = 2.08 \times 10^{-6} \text{ F}$.

13.   (a) $I_o = \dfrac{\varepsilon}{R} = \dfrac{20 \text{ V}}{5 \ \Omega} = 4.0 \text{ A}$.

(b) $\tau_L = \dfrac{L}{R} = \dfrac{4.0 \times 10^{-3} \text{ H}}{5.0 \ \Omega} = 8.0 \times 10^{-4} \text{ s}$. $I(4.0 \times 10^{-4} \text{ s}) = I_o \, e^{-t/\tau_L} = (4.0 \text{ A}) \, e^{-0.5} = 2.4 \text{ A}$.

(c) $V_L(4.0 \times 10^{-4} \text{ s}) = \varepsilon \, e^{-t/\tau_L} = (20 \text{ V}) \, e^{-0.5} = 12.1 \text{ V}$, $V_R = IR = (2.43 \text{ A})(5.0 \ \Omega) = 12 \text{ V}$.

19.   (a) $I_1 = I_2 = \dfrac{\varepsilon}{R_1 + R_2} = \dfrac{50 \text{ V}}{30 \ \Omega} = 1.67 \text{ A}$.

(b) Now L is a short, R_2 and R_3 are parallel. $I_1 = \dfrac{\varepsilon}{R_1 + R_2 R_3/(R_2 + R_3)} = \dfrac{50 \text{ V}}{10 \ \Omega + 12 \ \Omega} = 2.27 \text{ A}$.

$I_2 = \dfrac{R_3}{R_2 + R_3} I_1 = \dfrac{30 \ \Omega}{50 \ \Omega} (2.27 \text{ A}) = 1.36 \text{ A}$.

(c) Just before S is opened $I_L = I_1 - I_2 = 0.91 \text{ A}$. The falling magnetic flux in L will try to maintain this

current, so $I_2 = -I_L = -0.91 \text{ A}$, $I_1 = 0$.

(d) $I_1 = I_2 = 0$.

25. $\omega = \dfrac{1}{\sqrt{LC}} = \dfrac{1}{\sqrt{(0.20 \times 10^{-3} \text{ H})(5.0 \times 10^{-12} \text{ F})}} = 3.16 \times 10^7 \text{ rad/s}$.

31. $C = \dfrac{1}{4\pi^2 f^2 L}$, $f_1 = 540 \text{ Hz}$. $C_1 = \dfrac{1}{4\pi^2 (5.40 \times 10^5 \text{ Hz})^2 (2.5 \times 10^{-3} \text{ H})} = 3.48 \times 10^{-11} \text{ F}$.

$f_2 = 1600 \text{ Hz}$, $C_2 = \dfrac{1}{4\pi^2 (1.60 \times 10^6 \text{ Hz})^2 (2.5 \times 10^{-3} \text{ H})} = 3.96 \times 10^{-12} \text{ F}$.

37.   (a) At t = 0, all capacitors are shorts.

$\dfrac{1}{R_{eq}} = \dfrac{1}{R} + \dfrac{1}{3R/2}$, $R_{eq} = 0.60 \text{ R}$.

$I = \dfrac{\varepsilon}{R_{eq}} = \dfrac{\varepsilon}{0.60R} = 1.67 \dfrac{\varepsilon}{R}$.

(b) $I = \dfrac{\varepsilon}{R}$.

(c) $V = \varepsilon$ for both capacitors since no current through series resistors, $Q = C\varepsilon$.

Chapter 34 Alternating Current Circuits

1.　　$V(t) = V_o \sin\omega t = (12\ V)\sin(400\pi t)$.

7.　　$X_C = \dfrac{1}{2\pi fC} = X_L = 2\pi fL$,　　　$C = \dfrac{1}{4\pi^2 f^2 L} = \dfrac{1}{4\pi^2(10^3\ Hz)^2(5.0\times10^{-3}\ H)} = 5.07\times10^{-6}\ F$.

13.　　(a) $X_L = 2\pi fL = 2\pi(2.0\times10^4\ Hz)(20\times10^{-3}\ H) = 2.51\times10^3\ \Omega$.　　$I_o = \dfrac{V_o}{X_L} = \dfrac{9.0\ V}{2.51\times10^3\ \Omega} = 3.58\times10^{-3}\ A$.

　　　　(b) $X_L = 2\pi(60\ Hz)(20\times10^{-3}\ H) = 7.54\ \Omega$,　　　$I_o = \dfrac{9.0\ V}{7.54\ \Omega} = 1.19\ A$.

19.　　(a) $X_C = \dfrac{1}{2\pi fC} = \dfrac{1}{2\pi(500\ Hz)(2.0\times10^{-6}\ F)} = 159\ \Omega$.　　$X_L = 2\pi fL = 2\pi(500\ Hz)(0.20\ H) = 628\ \Omega$.

　　　　$Z = \sqrt{R^2 + (X_L - X_C)^2} = \sqrt{(500\ \Omega)^2 + (469\ \Omega)^2} = 686\ \Omega$.

　　　　(b) $I_o = \dfrac{V_o}{Z} = \dfrac{100\ V}{686\ \Omega} = 0.146\ A$.

　　　　(c) $\tan\phi = \dfrac{X_L - X_C}{R} = \dfrac{469\ \Omega}{500\ \Omega} = 0.938$,　　　$\phi = 43.2° = 0.753\ rad$.

　　　　$I(t) = I_o \sin(\omega t - \phi) = (0.146\ A)\sin(1000\pi t - 0.753)$.

　　　　(d) $X_C = \dfrac{1}{2\pi(500\ Hz)(2.0\times10^{-7}\ F)} = 1592\ \Omega$.　　$Z = \sqrt{(500\ \Omega)^2 + (964)^2} = 1086\ \Omega$.

　　　　$\tan\phi = -1.93$,　　　$\phi = -62.6° = -1.09\ rad$.　　　$I_o = \dfrac{100\ V}{1086\ \Omega} = 0.092\ A$.

　　　　$I(t) = (0.092\ A)\sin(1000\pi t + 1.09)$.

25.　　$V_{rms} = \dfrac{V_o}{\sqrt{2}} = \dfrac{50\ V}{\sqrt{2}} = 35.4\ V$.

31.　　$\omega_o = \dfrac{1}{\sqrt{LC}} = \dfrac{1}{\sqrt{(75\times10^{-3}\ H)(4.0\times10^{-6}\ F)}} = 1.83\times10^3\ rad/s$　independent of R.

37.　　$C_1 = \dfrac{1}{4\pi^2 f_o^2 L} = \dfrac{1}{4\pi^2(100\ Hz)^2(20.0\ H)} = 1.27\times10^{-7}\ F$.

　　　　$R = \dfrac{2\pi f_o L}{Q} = \dfrac{2\pi(100\ Hz)(20\ H)}{10} = 1257\ \Omega$.　　　$\Delta R = 1257\ \Omega - 200\ \Omega = 1057\ \Omega$.

43.　　(a) $P = I_p V_p = I_s V_s$,　　　$I_s = \dfrac{I_p V_p}{V_s} = \dfrac{(100\ A)(15\times10^3\ V)}{150\times10^3\ V} = 10\ A$.

　　　　(b) $\dfrac{P}{l} = I^2\left(\dfrac{R}{l}\right) = (10\ A)^2(8.6\times10^{-8}\ /m) = 8.6\times10^{-6}\ W/m$.

　　　　(c) $\dfrac{P}{l} = (100\ A)^2(8.6\times10^{-8}\ \Omega/m) = 8.6\times10^{-4}\ W/m$.

49. $A_V = -\dfrac{R_f}{R_i},$ So, $R_f = |A_V|R_i = (10)(250\ \Omega) = 2.50\times10^3\ \Omega.$

55. $X_C = \dfrac{1}{2\pi fC} = \dfrac{1}{2\pi(60\ \text{Hz})(2.0\times10^{-4}\ \text{F})} = 13.3\ \Omega,$ $X_L = 2\pi fL = 2\pi(60\ \text{Hz})(0.120\ \text{H}) = 45.2\ \Omega.$

$R_s = 70\ \Omega,$ $Z = \sqrt{R^2 + (X_L - X_C)^2} = \sqrt{(70\ )^2 + (31.9\ \Omega)^2} = 76.9\ \Omega.$

$I_{rms} = \dfrac{V_{rms}}{Z} = \dfrac{240\ \text{V}}{76.9\ \Omega} = 3.12\ \text{A}.$

(a) $V_R = I_{rms}\,R = (3.12\ \text{A})(50\ \Omega) = 156\ \text{V}.$ (b) $V_C = I_{rms}\,X_C = (3.12\ \text{A})(13.3\ \Omega) = 42\ \text{V}.$

(c) $Z_L = \sqrt{X_L^2 + R_L^2} = \sqrt{(45.2\ \Omega)^2 + (20\ \Omega)^2} = 49.4\ \Omega.$ $V_L = I_{rms}\,Z_L = (3.12\ \text{A})(49.4\ \Omega) = 154\ \text{V}.$

Chapter 35 Electromagnetic Waves and the Nature of Light

1. $f = \dfrac{c}{\lambda} = \dfrac{2.998\times10^8\ \text{m/s}}{3.25\times10^{-2}\ \text{m}} = 9.22\times10^9\ \text{Hz}.$

7.   (a) $E_o = cB_o = (2.998\times10^8\ \text{m/s})(2.15\times10^{-11}\ \text{T}) = 6.45\times10^{-3}\ \text{V/m}.$ (b) $\lambda = \dfrac{c}{f} = \dfrac{2.998\times10^8\ \text{m/s}}{760\times10^3\ \text{Hz}} = 394\ \text{m}.$

13. $\dfrac{E}{l} = \dfrac{\text{Power}}{c} = \dfrac{2.5\times10^{-3}\ \text{W}}{2.998\times10^8\ \text{m/s}} = 8.3\times10^{-12}\ \text{J/m}.$

19.   (a) $F_{av} = P_r\dfrac{A}{2} + P_a\dfrac{A}{2} = \left(\dfrac{2S_{av}}{c}\right)\dfrac{A}{2} + \left(\dfrac{S_{av}}{c}\right)\dfrac{A}{2} = \dfrac{3(250\ \text{W/m}^2)}{2.998\times10^8\ \text{m/s}}\dfrac{(0.25\ \text{m})^2}{2} = 7.8\times10^{-8}\ \text{N}.$

(b) $\Delta E = S_{av}\dfrac{A}{2}\Delta t = \dfrac{(250\ \text{W/m}^2)(0.25\ \text{m})^2}{2}(300\ \text{s}) = 2.3\times10^3\ \text{J}.$

(c) $\Delta T = \dfrac{\Delta E}{MC_{Al}} = \dfrac{\Delta E}{\rho_{Al}\,V_{Al}\,C_{Al}} = \dfrac{2.34\times10^3\ \text{J}}{(2.70\times10^3\ \text{kg/m}^3)(0.25\ \text{m})^2(0.002\ \text{m})(900\ \text{J/kg}\cdot\text{C}^\circ)} = 7.7^\circ\text{C}.$

25. $\theta_1 = \theta'_1 = 70^\circ$ to normal. $\sin\theta_2 = \dfrac{n_1\,\sin\theta_1}{n_2} = \dfrac{(1.00)\sin70^\circ}{1.33} = 0.707,\ \ \theta_2 = \sin^{-1}(0.707) = 45.0^\circ.$

31.   (a) $\sin\theta_2 = \dfrac{n_1\,\sin\theta_1}{n_2} = \dfrac{(1.7)\sin40^\circ}{1.4} = 0.781,$ $\theta_2 = 51.3^\circ.$

(b) $\sin\theta_1 = \dfrac{n_2\,\sin\theta_2}{n_1} = \dfrac{(1.7)\sin40^\circ}{1.4} = 0.781,$ $\theta_1 = 51.3^\circ.$

37. $n_2 = \dfrac{1}{\sin\theta_c} = \dfrac{1}{\sin43^\circ} = 1.47.$

43. $I_T = I_M \cos^2\theta,$ $\cos^2\theta = \dfrac{I_T}{I_M} = 0.25,$ $\theta = \cos^{-1}(\sqrt{0.25}\,) = 60°.$

49. $\theta = \tan^{-1}\!\left(\dfrac{1.33}{1.00}\right) = 53.1°.$

55. $n\sin\dfrac{\phi}{2} = \sin\!\left(\dfrac{\alpha+\phi}{2}\right),$ $\alpha = 2\sin^{-1}\!\left(n\sin\dfrac{\phi}{2}\right) - \varphi = 2\sin^{-1}(1.50\sin10°) - 20° = 10.2°.$

61. $\mathbf{E_1} = E_{1o}\cos(kx - \omega t)\,\mathbf{j},$ $\mathbf{E_2} = E_{2o}\cos\!\left(kx - \omega t - \dfrac{\pi}{2}\right)\mathbf{k} = -E_{2o}\sin(kx - \omega t)\,\mathbf{k},$

So, $\dfrac{E^2_1}{E^2_{1o}} + \dfrac{E^2_2}{E^2_{2o}} = \cos^2(kx - \omega t) + \sin^2(kx - \omega t) = 1.$

This is an equation for an ellipse in y-z plane with E_{1o} and E_{2o} the lengths of its semi-major and semi-minor axes.   So $\mathbf{E} = \mathbf{E_1} + \mathbf{E_2}$ is a vector that rotates clockwise in the y-z plane with the tip of $\mathbf{E}$ making an ellipse with semi–axes of lengths E_{1o} and E_{2o}. If $\phi \neq \dfrac{\pi}{2}$ then we have an ellipse whose major axis is rotated with respect to the y axis.

Chapter 36 Formation of Images

1.   (a) i = –o = –20.0 cm (behind mirror).   (b) d = o + |i| = 40.0 cm.

6.   So 31 in from the floor.

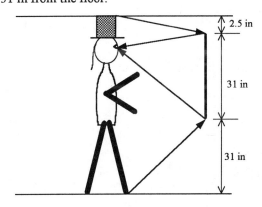

7. Top of mirror should be 2.5 in above her eye level so it is 5.5 in too low. She cannot see the upper 11 in of her body so she'll see $1 - \frac{11}{67} = \frac{56}{67}$ of her body.

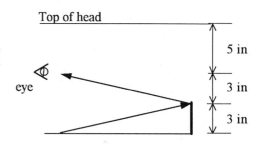

Top of head

5 in

3 in

3 in

eye

13. $\frac{1}{o} + \frac{1}{i} = \frac{1}{f} = \frac{1}{80 \text{ cm}} + \frac{1}{-32 \text{ cm}} = -0.0187 \text{ /cm},$ $f = -53.3 \text{ cm}.$

19. (a) o = 40 cm,   i = 40 cm   (b) o = 10 cm,   i = −20 cm,
     m = −1,   image size = 2.0 cm.   m = 2,   image size = 2.0 cm.

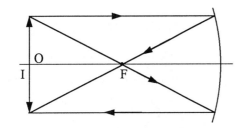

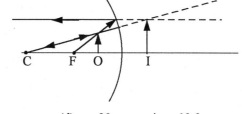

(c) o = 30 cm,   i = −15 cm,   (d) o = 20 cm,   i = −13.3 cm,
     $m = \frac{1}{2}$,   image size = 0.75 cm.   m = 0.667,   image size = 1.3 cm.

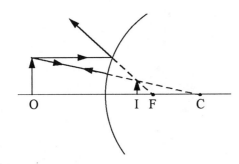

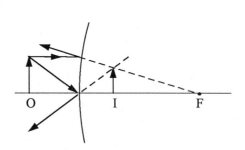

25. (a) $f = \frac{R}{2} = 40 \text{ cm}.$ $\frac{1}{o} = \frac{1}{f} - \frac{1}{i} = \frac{1}{40 \text{ cm}} - \frac{1}{400 \text{ cm}} = \frac{9}{400 \text{ cm}},$ o = 44.4 cm.

(b) $h' = mh = \left(\frac{-i}{o}\right)h = \left(\frac{-400 \text{ cm}}{44.4 \text{ cm}}\right)(2.0 \text{ cm}) = -18.0 \text{ cm}.$ (c) Image is inverted.

31. Air:
    (a) $\frac{1.5}{i} = \frac{(1.5 - 1.0)}{-15 \text{ cm}} - \frac{1.0}{10 \text{ cm}} = -0.133 \text{ /cm},$ $i = \frac{1.5}{-0.133 \text{ /cm}} = -11.3 \text{ cm}.$

    $m = -\frac{(1.0)(-11.25 \text{ cm})}{(1.5)(10 \text{ cm})} = 0.75.$

(b) $\dfrac{1.5}{i} = \dfrac{(1.5-1.0)}{-15\ \text{cm}} - \dfrac{1.0}{15\ \text{cm}} = -0.100\ /\text{cm},$ $\qquad i = \dfrac{1.5}{-0.100\ /\text{cm}} = -15\ \text{cm}.$

$m = -\dfrac{(1.0)(-15\ \text{cm})}{(1.5)(15\ \text{cm})} = 0.67.$

(c) $\dfrac{1.5}{i} = \dfrac{(1.5-1.0)}{-15\ \text{cm}} - \dfrac{1.0}{40\ \text{cm}} = -0.0583\ /\text{cm},$ $\qquad i = \dfrac{1.5}{-0.0583\ /\text{cm}} = -25.7\ \text{cm}.$

$m = -\dfrac{(1.0)(-25.7\ \text{cm})}{(1.5)(40\ \text{cm})} = 0.43.$

Water:

(a) $\dfrac{1.5}{i} = \dfrac{(1.5-1.33)}{-15\ \text{cm}} - \dfrac{1.33}{10\ \text{cm}} = -0.144\ /\text{cm},$ $\qquad i = \dfrac{1.5}{-0.144\ /\text{cm}} = -10.4\ \text{cm}.$

$m = -\dfrac{(1.33)(-10.4\ \text{cm})}{(1.5)(10\ \text{cm})} = 0.92.$

(b) $\dfrac{1.5}{i} = \dfrac{(1.5-1.33)}{-15\ \text{cm}} - \dfrac{1.33}{15\ \text{cm}} = -0.100\ /\text{cm},$ $\qquad i = \dfrac{1.5}{-0.100\ /\text{cm}} = -15\ \text{cm}.$

$m = -\dfrac{(1.33)(-15\ \text{cm})}{(1.5)(15\ \text{cm})} = 0.89.$

(c) $\dfrac{1.5}{i} = \dfrac{(1.5-1.33)}{-15\ \text{cm}} - \dfrac{1.33}{40\ \text{cm}} = -0.0446\ /\text{cm},$ $\qquad i = \dfrac{1.5}{-0.0446\ /\text{cm}} = -33.6\ \text{cm}.$

$m = -\dfrac{(1.33)(-33.6\ \text{cm})}{(1.5)(40\ \text{cm})} = 0.75.$

37. Top $\quad i_1 = -\dfrac{n_2}{n_1}\,d = \dfrac{-d}{1.33},$

Bottom, $i_2 = -\dfrac{(d+25)}{1.33}.$

Thickness $= i_1 - i_2 = \dfrac{25\ \text{cm}}{1.33} = 18.8\ \text{cm}.$

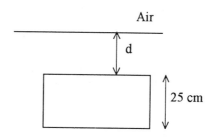

43. $\quad n = \dfrac{1}{\sin\theta_c} = \dfrac{1}{\sin 45°} = 1.41.$ $\qquad\qquad i = -\dfrac{n_2}{n_1}\,o = -\dfrac{1.00}{1.41}\,(2.0\ \text{cm}) = -1.41\ \text{cm}.$

48. $\quad m = -\dfrac{i}{o},$ $\qquad \dfrac{1}{o} + \dfrac{1}{i} = \dfrac{1}{f},$ $\qquad \dfrac{1}{o} + \dfrac{1}{-mo} = \dfrac{1}{f},$ $\qquad$ So, $\qquad \dfrac{m-1}{mo} = \dfrac{1}{f},$

and $\qquad o = d = \dfrac{(m-1)f}{m}.$

49. From Problem 36-48, $\qquad o = \dfrac{(m-1)}{m}\,f,$ $\qquad o_1 = \dfrac{m_1-1}{m_1}\,f_1 = o_2 = \dfrac{m_2-1}{m_2}\,f_2.$

$\dfrac{(1/3)-1}{1/3}\,(-f) = \dfrac{m_2-1}{m_2}\,(f),$ $\qquad 2f = \dfrac{m_2-1}{m_2}\,f,$ $\qquad$ So, $\qquad m_2 = -1.$

Chapter 37 Lenses and Optical Instruments

1.   First surface (air to glass), $\dfrac{n_2}{i_1} = \dfrac{n_2 - n_1}{R_1} - \dfrac{n_1}{o}$,

So, $\dfrac{1.50}{i_1} = \dfrac{(1.50 - 1.00)}{20 \text{ cm}} - \dfrac{1.00}{15 \text{ cm}} = -0.0417 \text{ /cm},$ $i_1 = -36.0$ cm.

Second surface (glass to air), $o_2 = 2R - i_1 = 40 \text{ cm} - (-36.0 \text{ cm}) = 76.0$ cm.

$\dfrac{n_2}{i_2} = \dfrac{(n_2 - n_1)}{R_2} - \dfrac{n_1}{o_2} = \dfrac{1.00}{i_2} = \dfrac{(1.00 - 1.50)}{-20 \text{ cm}} - \dfrac{1.50}{76.0 \text{ cm}} = 0.00526 \text{ /cm},$

$i_2 = 190$ cm to right of sphere.

7.   (a) $\dfrac{1}{f} = (1.5 - 1)\left(\dfrac{1}{20} - \dfrac{1}{-20}\right) = \dfrac{0.5}{10}$,   (b) $\dfrac{1}{f} = (1.5 - 1)\left(\dfrac{1}{20} - \dfrac{1}{-10}\right) = \dfrac{1.5}{20}$

     $f = 20$ cm. $f = 13.3$ cm.

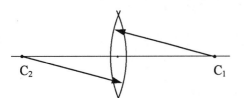   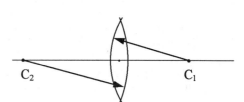

(c) $\dfrac{1}{f} = (1.5 - 1)\left(\dfrac{1}{20} - \dfrac{1}{\infty}\right) = \dfrac{1}{40}$ (d) $\dfrac{1}{f} = (1.5 - 1)\left(\dfrac{1}{-20} - \dfrac{1}{20}\right) = \dfrac{-0.5}{10}$

     $f = 40$ cm. $f = -20$ cm.

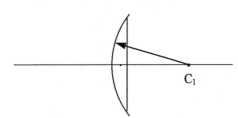   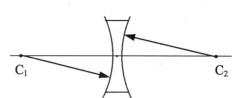

(e) $\dfrac{1}{f} = (1.5 - 1)\left(\dfrac{1}{-20} - \dfrac{1}{\infty}\right) = \dfrac{-0.5}{20}$ (f) $\dfrac{1}{f} = (1.5 - 1)\left(\dfrac{1}{-10} - \dfrac{1}{-20}\right) = \dfrac{0.5}{-20}$

     $f = -40$ cm. $f = -40$ cm.

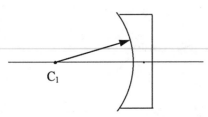   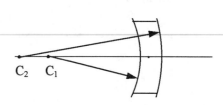

(g) $\dfrac{1}{f} = (1.5 - 1)\left(\dfrac{1}{-20} - \dfrac{1}{-10}\right) = \dfrac{0.5}{20}$, $f = 40$ cm.

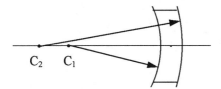

9. $\dfrac{n'}{o} + \dfrac{n}{i_1} = \dfrac{n - n'}{R_1}$,   and $\dfrac{n}{-i_1} + \dfrac{n'}{i} = \dfrac{n' - n}{R_2}$.   Add these two and

$n'\left(\dfrac{1}{o} + \dfrac{1}{i}\right) = (n - n')\left(\dfrac{1}{R_1} - \dfrac{1}{R_2}\right)$.   So, $\dfrac{1}{o} + \dfrac{1}{i} = \left(\dfrac{n}{n'} - 1\right)\left(\dfrac{1}{R_1} - \dfrac{1}{R_2}\right)$.

13. From Problem 37-9: $\dfrac{1}{f_w} = \left(\dfrac{n}{n'} - 1\right)\left(\dfrac{1}{R_1} - \dfrac{1}{R_2}\right)$, in water, $\dfrac{1}{f_a} = (n - 1)\left(\dfrac{1}{R_1} - \dfrac{1}{R_2}\right)$, in air.

$\dfrac{f_a}{f_w} = \dfrac{(n/n') - 1}{n - 1} = \dfrac{(1.50/1.33) - 1}{1.50 - 1} = 0.256$. $f_a = 0.256 f_w = (0.256)(25\ \text{cm}) = 6.4$ cm.

Image is 6.4 cm from lens.

19. (a) $\dfrac{1}{f} = \left(\dfrac{n}{n'} - 1\right)\left(\dfrac{1}{R_1} - \dfrac{1}{R_2}\right) = \left(\dfrac{1.3}{1.8} - 1\right)\left(\dfrac{1}{-20\ \text{cm}} - \dfrac{1}{20\ \text{cm}}\right) = 0.0278\ /\text{cm}$, $f = 36$ cm.

(b) For real image to form, the object must be outside the focal point, or $o > 36$ cm.

25. $\dfrac{1}{25} + \dfrac{1}{i} = \dfrac{1}{f} = \dfrac{1}{21} + \dfrac{1}{(i + 3)}$, $\dfrac{(i + 25)}{25i} = \dfrac{(i + 3 + 21)}{21(i + 3)}$, $4i^2 + 12i - 1575 = 0$,

$i = -\dfrac{3}{2} \pm 19.90$, $i = 18.4$ cm, $f = 10.6$ cm.

31. First lens, $o_1 = \infty$, $i_1 = f_1 = 6.0$ cm.

Second lens, $o_2 = 3.0 - i_1 = -3.0$ cm, $\dfrac{1}{i_2} = \dfrac{1}{f_2} - \dfrac{1}{o_2} = \dfrac{1}{6.0\ \text{cm}} - \dfrac{1}{-3.0\ \text{cm}} = 0.50\ /\text{cm}$,

$i_2 = 2.0$ cm.   Focal points are 2.0 cm from each lens on the outside of the combination.

37. (a) $o = \infty$, $i = -1.48$ m, $f = -148$ cm.

(b) $o = 23$ cm, $i = -73$ cm, $\dfrac{1}{f} = \dfrac{1}{23} + \dfrac{1}{-73} = 0.0298\ /\text{cm}$, $f = 33.6$ cm.

43. $f_o = \dfrac{(-25\ \text{cm})L}{M f_o} = \dfrac{(-25\ \text{cm})(18\ \text{cm})}{(-1000)(0.35\ \text{m})} = 1.3$ cm.

49. $\theta = \dfrac{D_o}{f_o} = \dfrac{D_e}{f_e}$,

$D_e = D_o \dfrac{f_e}{f_o} = \dfrac{D_o}{M} = \dfrac{8.0 \text{ cm}}{40} = 0.20 \text{ cm.}$

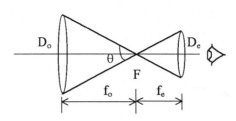

55. First surface, $R_1 = \infty$, $\dfrac{n}{i_1} = \dfrac{-1}{d}$, or, $i_1 = -nd$.

Second surface, $R_2 = \infty$, $o_2 = x - i_1 = x + nd$, $\dfrac{1}{i_2} = -\dfrac{n}{o_2} = -\dfrac{n}{x + nd}$.

$i_2 = -\left(d + \dfrac{x}{n}\right)$, which approaches $-d$ as $x \to 0$.

61. $\theta_1 \text{ (game)} = \dfrac{2.0 \text{ m}}{80 \text{ m}} = 0.025 \text{ rad,}$ $\theta_2 \text{ (TV)} = \dfrac{15 \text{ cm}}{300 \text{ cm}} = 0.050 \text{ rad.}$

The television viewer sees an image twice as large as the spectator.

Chapter 38 Interference and Diffraction

1. $\sin\theta \approx \theta = \dfrac{m\lambda}{d}$, $\theta_1 = \dfrac{600\times10^{-9} \text{ m}}{0.12\times10^{-3} \text{ m}} = 0.0050 \text{ rad,}$ $\theta_3 = \dfrac{3(600\times10^{-9} \text{ m})}{0.12\times10^{-3} \text{ m}} = 0.015 \text{ rad.}$

7. $\lambda = \dfrac{dy}{mD} = \dfrac{(0.15\times10^{-3} \text{ m})(0.028 \text{ m})}{5(1.5 \text{ m})} = 5.60\times10^{-7} \text{ m.}$

13. $E_R = \dfrac{E_o \sin N\alpha}{\sin\alpha} = \dfrac{E_o \sin(N\phi/2)}{\sin(\phi/2)} = \dfrac{E_o \sin4\phi}{\sin(\phi/2)} = 0$,

Require $4\phi = \pi$ or $\phi = \dfrac{\pi}{4}$.

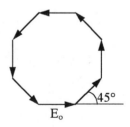

19. $\lambda_o = \dfrac{2nt}{m} = \dfrac{2(1.40)(250\times10^{-9} \text{ m})}{m} = \dfrac{7.00\times10^{-7}}{m}$, $m = 1$, $\lambda_o = 700 \text{ nm (Red).}$

25. $\lambda_n = \dfrac{\lambda_o}{n} = \dfrac{2t}{(m + 1/2)}$, In 10 cm we have 180 fringes or $m = 179$,

$\dfrac{\lambda_o}{n} = \dfrac{2(4.11\times10^{-5} \text{ m})}{179 + 1/2} = 4.58\times10^{-7} \text{ m,}$ $n = \dfrac{\lambda_o}{4.58\times10^{-7} \text{ m}} = \dfrac{589\times10^{-9} \text{ m}}{4.58\times10^{-7} \text{ m}} = 1.29.$

31. (a) $\Delta t = \dfrac{1}{\Delta f} = \dfrac{1}{3.0 \times 10^6 \text{ Hz}} = 3.33 \times 10^{-7}$ s.

(b) $c\Delta t = (3.00 \times 10^8 \text{ m/s})(3.33 \times 10^{-7} \text{ s}) = 100$ m.

(c) No, since $c\Delta t$ is large compared to the path difference between the two slits.

37. $R = \dfrac{y}{\tan\theta} = \dfrac{2.5 \times 10^{-3} \text{ m}}{\tan(0.229°)} = 0.625$ m.

43. $\theta_1 = \sin^{-1}\left(\dfrac{\lambda}{b}\right) \approx \dfrac{\lambda}{b} = \dfrac{5.0 \times 10^{-7} \text{ m}}{0.16 \times 10^{-3} \text{ m}} = 3.125 \times 10^{-3}$ rad, $\qquad \theta_2 = \dfrac{2\lambda}{b} = \dfrac{2(5.0 \times 10^{-7} \text{ m})}{0.16 \times 10^{-3} \text{ m}} = 6.25 \times 10^{-3}$ rad.

$D = \dfrac{\Delta y}{\Delta \theta} = \dfrac{6.0 \times 10^{-3} \text{ m}}{3.125 \times 10^{-3} \text{ rad}} = 1.9$ m.

48. The fifth minimum of the interference term $\cos^2\alpha$ must coincide with the first minimum of the diffraction term $\left(\dfrac{\sin\beta}{\beta}\right)^2$. The fifth minimum occurs at $\alpha = \dfrac{9\pi}{2}$ and the first minimum occurs at $\beta = \pi$.

Since $\alpha = \dfrac{\pi d \sin\theta}{\lambda}$, $\qquad$ and $\qquad \beta = \dfrac{\pi b \sin\theta}{\lambda}$, $\qquad \dfrac{d}{b} = \dfrac{\alpha}{\beta} = 4.5$.

49. From Problem 38-48, $\qquad n = \dfrac{2d}{b} = 2\left(\dfrac{4b}{b}\right) = 8$.

55. (a) $\theta(\text{microscope}) = \dfrac{1.22\lambda}{D} = \dfrac{s}{R} = \dfrac{s}{f_o}$, $\qquad s = \dfrac{1.22\lambda f_o}{D}$.

(b) $D = 2f_o \tan\alpha$, $\qquad s = \dfrac{1.22\lambda f_o}{2f_o \tan\alpha} = \dfrac{0.61\lambda}{\tan\alpha}$.

(c) $s = \dfrac{(0.61)(550 \times 10^{-9} \text{ m})}{\tan 60°} = 1.94 \times 10^{-7}$ m.

61. (a) $m = \dfrac{d\sin\theta}{\lambda} = \dfrac{(0.01 \text{ m})(\sin 90°)/5000}{570.72 \times 10^{-9} \text{ m}} = 3.5$, $\qquad$ maximum $\qquad m = 3$.

(b) $N = \dfrac{\bar{\lambda}}{m\Delta\lambda} = \dfrac{570.25 \times 10^{-9} \text{ m}}{(3)(0.010 \times 10^{-9} \text{ m})} = 1.90 \times 10^4$, $\qquad$ So, $\qquad l = \dfrac{1.90 \times 10^4}{5000 \text{ /cm}} = 3.80$ cm.

67. $\dfrac{I}{4I_o} = \cos^2\left(\dfrac{\pi d y}{\lambda D}\right)$, $\qquad$ and $\qquad \dfrac{\pi d y_3}{\lambda D} = 3\pi$,

So, $\qquad y_3 = \dfrac{3\lambda D}{d} = 1.80 \times 10^{-3}$ m $\left(\dfrac{\lambda D}{d} = 0.60 \times 10^{-3} \text{ m}\right)$.

$\dfrac{I}{4I_o} = \dfrac{1}{3}$, $\qquad \Rightarrow \qquad \cos\dfrac{\pi d y}{\lambda D} = \pm\sqrt{\dfrac{1}{3}}$, $\qquad$ and $\qquad \dfrac{\pi r y}{\lambda D} = (3\pi \pm 0.9553)$ rad.

So, $\qquad y = \dfrac{\lambda D}{\pi d}(\pi \pm 0.9553) = \dfrac{0.60 \times 10^{-3} \text{ m}}{\pi}(3\pi \pm 0.9553) = (1.80 \pm 0.18) \times 10^{-3}$ m

$= 1.62 \times 10^{-3}$ m and 1.98×10^{-3} m.

73. (a) $b\sin\theta = m\lambda$, $\qquad \theta = \pm\sin^{-1}\left(\dfrac{m\lambda}{b}\right) = \pm\sin^{-1}\left[\dfrac{(1)(10\ mm)}{25\ mm}\right] = \pm23.6°$.

(b) $\sin\left[\dfrac{(1)(30\ mm)}{25\ mm}\right] > 1$, $\qquad$ there would be no minima.

Chapter 39 The Special Theory of Relativity

1. $\Delta t = \dfrac{\Delta t'}{\sqrt{1 - u^2/c^2}} = \dfrac{\Delta t'}{\sqrt{1 - (0.90c/c)^2}}$, $\qquad$ So, $\qquad \dfrac{\Delta t}{\Delta t'} = 2.29$.

7. (a) $\Delta t = \dfrac{\Delta t'}{\sqrt{1 - u^2/c^2}} = \dfrac{4.0\ \mu s}{\sqrt{1 - (0.60)^2}} = 5.0\ \mu s$.

(b) $\Delta x = \dfrac{u\Delta t'}{\sqrt{1 - u^2/c^2}} = \dfrac{(0.60)(3.00\times10^8\ m/s)(4.0\times10^{-6}\ s)}{\sqrt{1 - (0.60)^2}} = 900\ m$.

13.

u/c	0.1	0.5	0.6	0.7	0.8	0.9	0.95	0.97	0.98	0.99	0.995
L/L'	0.995	0.866	0.800	0.714	0.600	0.436	0.314	0.243	0.191	0.141	0.0999

19. $t = 12\ yr = \dfrac{(9.0\ yr)\sqrt{1 - u^2/c^2}}{(u/c)}$, $\qquad (144)\left(\dfrac{u^2}{c^2}\right) = 81\left(1 - \dfrac{u^2}{c^2}\right)$,

So, $\qquad \dfrac{u}{c} = \sqrt{\dfrac{81}{225}} = 0.60$, and $\qquad u = 0.60c$.

25. $x' = \dfrac{x - ut}{\sqrt{1 - u^2/c^2}} = \dfrac{1000\ m - (0.80)(3.00\times10^8\ m/s)(4.0\times10^{-6}\ s)}{\sqrt{1 - (0.80)^2}} = 67\ m$.

$y' = y = 500\ m$, $\qquad z' = z = 500\ m$.

$t' = \dfrac{t - ux/c^2}{\sqrt{1 - u^2/c^2}} = \dfrac{4.0\times10^{-6}\ s - (0.80)(3.00\times10^8\ m/s)^{-1}(10^3\ m)}{\sqrt{1 - (0.80)^2}} = 2.2\times10^{-6}\ s$.

31. (a) $v_x = \dfrac{v'_x + u}{1 + uv'_x/c^2} = \dfrac{(0.70c + 0.50\ c)}{1 + (0.50c)(0.70c)/c^2} = 0.89c$.

(b) $v_x = \dfrac{-0.70c + 0.50c}{1 + (0.50c)(-0.70c)/c^2} = -0.31c$.

37. $v'_x = (0.95c)\cos60° = 0.475c$, $\qquad v'_y = (0.95c)\sin60° = 0.823c$.

$v_x = \dfrac{0.475c + 0.90c}{1 + (0.475c)(0.90c)/c^2} = 0.963c$, $\qquad v_y = \dfrac{(0.823c)\sqrt{1 - (0.90)^2}}{1 + (0.475c)(0.90c)/c^2} = 0.251c$.

$v = \sqrt{v_x^2 + v_y^2} = 0.995c$.

43. $mc^2 = 938.8$ MeV, $P = \dfrac{1}{c}\sqrt{E^2 - (mc^2)^2} = \dfrac{1}{c}\sqrt{(1500\text{ MeV})^2 - (938.8\text{ MeV})^2}$

$$= 1.17\times10^3 \text{ MeV/c} = 6.24\times10^{-19} \text{ kg·m/s.}$$

49.   (a) $E = (10^{-4})(25\text{ kg})(3.00\times10^8\text{ m/s})^2 = 2.3\times10^{14}$ J.

(b) Power $= \dfrac{2.25\times10^{14}\text{ J}}{1.5\times10^{-6}\text{ s}} = 1.5\times10^{20}$ W.

(c) Power $= \dfrac{8.0\times10^{19}\text{ J}}{(365)(24)(3600\text{ s})} = 2.5\times10^{12}$ W.

55.   (a) In S', $p'_\pi = 0$, $E' = mc^2 = 135$ MeV $= 2.16\times10^{-11}$ J.

$p'_1 = \dfrac{1.08\times10^{-11}\text{ J}}{3.00\times10^8\text{ m/s}} = 3.60\times10^{-20}$ kg·m/s,

$p'_2 = -3.60\times10^{-20}$ kg·m/s.

$E'_1 = \dfrac{E'}{2} = 1.08\times10^{-11}$ J, $E'_2 = 1.08\times10^{-11}$ J.

(b) In S, $p_\pi = \dfrac{p'_\pi + uE'_\pi/c^2}{\sqrt{1-\beta^2}} = \dfrac{0 + (0.90/c)(2.16\times10^{-11}\text{ J})}{\sqrt{1-0.90^2}} = 1.49\times10^{-19}$ kg·m/s.

$E_\pi = \dfrac{E'_\pi + up'_\pi}{\sqrt{1-\beta^2}} = \dfrac{2.16\times10^{-11}\text{ J}}{\sqrt{1-0.90^2}} = 4.95\times10^{-11}$ J.

$p_1 = \dfrac{p'_1 + (u/c^2)E'_1}{\sqrt{1-\beta^2}} = \dfrac{3.60\times10^{-20} + (0.90/c)(1.08\times10^{-11})}{\sqrt{1-0.90^2}} = 1.57\times10^{-19}$ kg·m/s.

$E_1 = \dfrac{E'_1 + up'_1}{\sqrt{1-\beta^2}} = \dfrac{1.08\times10^{-11} + (0.90c)(3.60\times10^{-20})}{\sqrt{1-0.90^2}} = 4.71\times10^{-11}$ J.

$p_2 = \dfrac{-3.60\times10^{-20} + (0.90/c)(1.08\times10^{-11})}{\sqrt{1-0.90^2}} = -8.26\times10^{-21}$ kg·m/s.

$E_2 = \dfrac{1.08\times10^{-11} + (0.90c)(-3.60\times10^{-20})}{\sqrt{1-0.90^2}} = 2.48\times10^{-12}$ J.

(c)   Before   After

net p $= 1.49\times10^{-19}$ kg·m/s,   net p $= 1.57\times10^{-19} - 8.26\times10^{-21} = 1.49\times10^{-19}$ kg·m/s,

net E $= 4.95\times10^{-11}$ J.   net E $= 4.71\times10^{-11} + 2.48\times10^{-12} = 4.96\times10^{-11}$ J.